Solution Manual

Fundamentals of Environmental Sampling and Analysis

Second Edition

Chunlong Zhang
University of Houston-Clear Lake
Texas, USA

Registered Office
John Wiley & Sons, Inc., 111 River Street, Hoboken, NJ 07030, USA

For details of our global editorial offices, customer services, and more information about Wiley products visit us at www.wiley.com.

Wiley also publishes its books in a variety of electronic formats and by print-on-demand. Some content that appears in standard print versions of this book may not be available in other formats.

Library of Congress Cataloging-in-Publication Data
Names: Zhang, Chunlong, 1964- author. | John Wiley & Sons, publisher.
Title: Fundamentals of environmental sampling and analysis / Chunlong Zhang.
Description: Second edition. | Hoboken, NJ : JW-Wiley, 2024. | Includes bibliographical references and index.
Identifiers: LCCN 2023057815 (print) | Set ISBN 9781394244621 | ISBN 9781394241651 (epdf) |
    ISBN 9781394241668 (epub) | ISBN 9781394241644 (SM)
Subjects: LCSH: Environmental sampling. | Environmental sciences--Statistical methods.
Classification: LCC GE45.S75 Z43 2024 (print) | LCC GE45.S75 (ebook) | DDC 628--dc23/eng/20240131
LC record available at https://lccn.loc.gov/2023057815
LC ebook record available at https://lccn.loc.gov/2023057816

Cover Design: Wiley
Cover Image: © Iana Kunitsa/Getty Images

Set in 9.5/12.5pt STIXTwoText by Integra Software Services Pvt. Ltd, Pondicherry, India

Chapter 1

Questions

1) **Give an example for each of the objectives of environmental sampling and analysis listed in the text.**
   Students' answers may vary. Corresponding to each listed objective, examples are:
   a) Sampling and analysis of contaminant concentrations in wastewater effluent from an industrial source to comply with effluent standards (e.g., NPDES).
   b) Monitor atmospheric ozone to determine air quality change over time against the NAAQ standard.
   c) Collections of samples and analysis of contaminant concentrations in the air surrounding an oil spill site to ensure health standards are met.
   d) Collections of soil and groundwater samples for the analysis of pollutants in a former gasoline station where a bioremediation project is underway. The information on how much contaminant we currently have, and how quickly is the progress of this ongoing project is important for us to make decision and take action in environmental remediation.

2) **Give examples of practice that will cause data to be scientifically defective or legally nondefensible.**
   Some of these examples are: improperly trained sampler and analyst, lack of good laboratory practice, knowingly falsifying test results, using nonvalidated equipment, unsecured chain of custody, nontraceable standards for instruments, misconduct, conflict of interest (e.g., contract labs oftentimes work for the company that hired them), ineffective ethics programs, etc.

3) **Define and give examples of systematic errors and random errors.**
   Determinate errors (systemic errors) produce a known bias in the data. These errors can be traced and corrected. They are avoidable mistakes that are known to have occurred or were found later. Measurements that result from these types of errors can be discarded. Random errors are indeterminate errors. They cannot be identified or compensated for, and statistics must be applied to deal with the data.

4) **Why are sampling and analysis integral parts of data quality? Between sampling and analysis, which one often generates more errors? Why?**
   If a sample isn't collected properly, then all subsequent careful lab work is useless. On the other hand, if an analyst is unable to define data quality (precision, accuracy), then such data are also useless and the money and time spent in collecting these samples are wasted. Most people

Fundamentals of Environmental Sampling and Analysis, Second Edition. Chunlong Zhang.
© 2024 John Wiley & Sons, Inc. Published 2024 by John Wiley & Sons, Inc.
Companion Website: www.wiley.com/go/EnvironmentalSamplingandAnalysis2e

would think that data quality is supported by a state-of-the-art instrument in the lab analysis, rather than sampling stage. However, this is typically not true. On the opposite, most error comes from sampling rather than lab analysis. For example, a sophisticated analytical instrument cannot justify the variations of chemicals in a very heterogeneous matrix.

5) **Describe how errors in environmental data acquisition can be minimized and quantified.**

Errors can be quantified and minimized through a quality assurance/quality control (QA/QC) program. QA is more of a management system ensuring QC is working properly, and QC is a system of technical activities to meet certain data specifications. The use of a QA/QC program and adequate protocols will reduce errors associated with sampling, sample preservation, sample transportation, sample preparation, and sample analysis.

6) **Why can't standard errors be added, but variances can?**

We cannot just add the standard errors or standard deviations, but we can add the variances as long as the variables are assumed to be independent such as errors from the stages of sampling and analysis. This is because during the variance calculations, the differences between every individual value from the mean are squared to get the positive values.

7) **How does environmental analysis differ from traditional analytical chemistry?**

Analytical chemists trained in traditional chemistry curriculum may not be immediately adapted to environmental analysis. Environmental analysis differs from traditional analytical chemistry in many ways, for example:

- Traditional analytical chemists stay in the lab, whereas environmental analytical chemists deal with measurements both *in situ* and in the lab.
- The analytical costs of environmental chemicals are typically high.
- There are always a large number of samples that require instrument automation.
- Sample matrices are often complex and unknown (water, air, soil, waste, and living organisms).
- The concentrations are typically very low. They are measured at ppm, ppb, ppt, or even lower levels.
- The markets and analytical protocols are driven by regulations. An environmental analyst needs a working knowledge of the regulations for the purposes of both regulatory compliance and regulatory enforcement.

8) **Describe the difference between "classical" and "modern" analysis.**

Modern analysis typically uses more or less sophisticated instrumental methods that make it possible to detect small quantities of almost anything. Examples of modern methods are spectrometric, electrometric, and chromatographic methods. The differences between "classical" and "modern" analysis are very arbitrary and ever-changing as technology advances. However, it is generally accepted that any "wet chemistry based" methods are termed "classical," while anything employing relatively sophisticated instrumentation is considered "modern" methods. Classical methods use wet chemicals analysis, mostly volumetric and gravimetric methods.

9) **Why should a QA/QC program be used in an environmental lab or environmental analysis consulting lab to ensure they meet guidelines?**

QA/QC programs are implemented not only to minimize errors from both sampling and analysis, but also to quantify the errors in the measurement. Knowing how much the error and the means to minimize the errors will ensure guidelines are met.

10) **Why might an analyst in the lab need to communicate well with the field sampler?**

    Students' answers may vary.

    Communication is always essential in a team project consisting of sampling and analysis. For example, the analyst needs to communicate to the field sampler for proper sample preservation and storage protocols if optimal results are to be achieved.

11) **Environmental professionals having strong sampling and analysis skills must not be timid in applying them to other types of sampling and analysis. Give an example of a discipline (e.g., pharmaceuticals, food and beverages, petroleum, forensics) or a scenario (e.g., detection of COVID-19) that you might pursue or encounter in your career. Elaborate your analysis with supporting publications.**

    Students' answers may vary. For example, as Hass indicated (Charles N. Haas, Coronavirus and Environmental Engineering Science, *Environ Engr Sci*, 37(4):233–234, 2020), in applying to the global COVID-19 pandemic, "environmental engineering and science researchers and practitioners must not be timid in applying them to effect restoration and improvement of the publics' health." While we already know about the fate, transport, and control of chemical and biological agents, we can apply them to deal with the spreading and control of COVID-19 virus. Specifically in applying sampling and analysis principles, as we will learn from this course, composite sampling was a strategy used to overcome the shortfall of testing kits available during the outbreak of COVID-19.

12) **A chemist is arguing that sampling is *not* as important as analysis. His concern is whether there is a need for a sampling course in an environmental curriculum. His main rationale is that most employers and governmental agencies already have their training courses alongside very specific and detailed procedures. Another consultant, on the contrary, argues that sampling should be given more weight than analysis. His main rationale is that a company always sends samples to commercial laboratories for analyses, and you do not become an analytical chemist by taking one course. For each of these two arguments, specify whether you agree or disagree and clearly state your supporting argument as to why you agree or disagree.**

    Answers to this question will vary, but the main theme of the discussions is to illustrate the importance of both sampling and analysis for the environmental professionals and the graduating students in the environmental curriculum.

    a) "Sampling is not as important as analysis," and "is there a need for a sampling course"? Sampling is just as important, or even more important than analysis, because that is where the error originates. In most cases, a second chance is not available to get the same original sample. The sampling procedure must be rigorous, ensuring that a perfect representative sample is collected and at no time is the sample or sample bottle contaminated by the collector. The need for a "sampling" component in the environmental curriculum is obvious. Most environmental programs emphasize the study of environmental issues rather than the scientific tools needed to investigate these issues (Clement, 1992, *Analytical Chemistry*, vol. 64, No. 22, 1076A–1081A). As Clement (1992) described, many students seldom realize the consequences of poor sampling, they just assume an analytical result generated by a million-dollar state-of-the-art instrument is a number "carved in stone." This is not the case in reality, many times, sampling determines the magnitude of errors in the entire data acquisition process.

b) Another consultant states: "Sampling should be given more weight than analysis," He says that his company always sends samples to commercial laboratories for analysis and you do not become an analytical chemist from one class. To some extent, the statement "Sampling should be given more weight than analysis" is supported by the fact that the majority of errors starts from sampling stages. Both sampling and the subsequent analysis of that sample are equally important and they are interdependent upon each other for obtaining good data. The importance of sampling is obvious. If the sample is not collected properly, it does not fully represent our system and all analysis is in vain! Once a representative sample is obtained, the results depend on the chemist who performs the analysis. If this person cannot define a level of analytical error and calculate precision, accuracy, and % recovery, the data is useless as well! Both field sampler and analyst are vitally important and good communication must be established between them in order to achieve "scientific reliability" in the data. The rest of argument from this consultant appears to be incorrect. First, we should realize that college courses are by no means designed to train students to become experts, but rather to equip them with the basic skills and knowledge for a given field. Even if a company does not do their samples on their own, the employees still have to communicate with the analytical company for the better outcomes of environmental data.

13) **For the results shown in Example 1.1, the volumetric device can read to 0.01 mL with the desired accuracy. For the mean and standard deviation of 9.86333 mg/L and 0.20474, respectively, tell why any of the following reporting is not correct or adequate: 9.86333, 9.86, 9.86 ± 0.01, 9.863 ± 0.205, 9.9 ± 0.2?**

- 9.86333: The uncertainty should be reported with the answer. Also, the answer has not been properly rounded off.
- 9.86: The uncertainty should be reported with the answer.
- 9.86 ± 0.01: The uncertainty is 0.20 not 0.01.
- 9.863 ± 0.205: It would be better to round off one more place.
- 9.9 ± 0.2: The rounding is overboard based on the accuracy of the volumetric device.

Problems

1) **Assume sampling and analysis are two independent processes, the sampling variance due to mercury concentrations in soil is 1.52 $(mg/kg)^2$, and the laboratory analysis variance for the measurement of Hg concentration is 0.032 $(mg/kg)^2$. (a) What is the sum of variance? (b) Calculate the standard deviation of the overall process. Report in the correct units.**

a) Applying Eq. 1.1: $s^2 = 1.52^2 + 0.032^2 = 2.311 \left(\dfrac{mg}{kg}\right)^2$

b) Applying Eq. 1.5: $SD = \sqrt{s^2} = \sqrt{2.311} = 1.52 \dfrac{mg}{kg}$

2) **If the percentage errors for a project surveying pesticide in soils during the sampling stage and analysis stage are known to be 80% and 20%, respectively, what is the overall error of measuring pesticide concentration?**

From Eq. 1.1, we have $s_t = \sqrt{s_s^2 + s_a^2} = \sqrt{0.80^2 + 0.20^2} = 0.8246$ or 82.46%.

3) **A solution made from the primary standard of $K_2Cr_2O_7$ contains 5.00 mg/L Cr. This solution was measured 5 times using a colorimetric method, and the measured concentrations are: 4.80, 5.13, 4.98, 4.83, 4.56 mg/L. Report the variance, standard deviation, and standard error of the mean.**

$$mean = \frac{4.80 + 5.13 + 4.98 + 4.83 + 4.56}{5} = 4.86 \frac{mg}{L}$$

Apply Eq. 1.4 for variance:

$$s^2 = \frac{(4.80 - 4.86)^2 + (5.13 - 4.86)^2 + (4.98 - 4.86)^2 + (4.83 - 4.86)^2 + (4.56 - 4.86)^2}{6 - 1}$$

$$= 0.04545 \left(\frac{mg}{L}\right)^2 \approx 0.05 \left(\frac{mg}{L}\right)^2$$

Applying Eq. 1.5 for standard deviation:

$$SD = \sqrt{0.04545} = 0.2139 \frac{mg}{L} \approx 0.21 \frac{mg}{L}$$

Applying Eq. 1.7 for SEM:

$$SEM = \frac{s}{\sqrt{n}} = \frac{0.2139}{\sqrt{5}} = 0.0953 \frac{mg}{L} \approx 0.10 \frac{mg}{L}$$

Alternatively, students can use Excel's function VAR () and STEDV () to quickly calculate variance and standard deviation, respectively (see Chapter 3, Table 3.2).

4) **A soil standard reference material has its certified lead (Pb) concentration of 17.3 mg/kg. The replicate analysis through an isotope dilution ICP-MS obtained the following results: 17.06, 16.90, 17.12, 17.01, 17.80, 17.16 mg/kg. Report the variance, standard deviation, and standard error of the mean.**

$$mean = \frac{17.06 + 16.90 + 17.12 + 17.01 + 17.80 + 17.16}{6} = 17.175 \frac{mg}{kg}$$

Apply Eq. 1.4 for variance:

$$s^2 = \frac{\begin{array}{c}(17.06 - 17.175)^2 + (16.90 - 17.175)^2 + (17.12 - 17.175)^2 + \\ (17.01 - 17.175)^2 + (17.80 - 17.175)^2 + (17.16 - 17.175)^2\end{array}}{6 - 1} = 0.10199 \left(\frac{mg}{kg}\right)^2 \approx 0.10 \left(\frac{mg}{kg}\right)^2$$

Applying Eq. 1.5 for standard deviation:

$$SD = \sqrt{0.10199} = 0.3194 \frac{mg}{kg} \approx 0.32 \frac{mg}{kg}$$

Applying Eq. 1.7 for SEM:

$$SEM = \frac{s}{\sqrt{n}} = \frac{0.3194}{\sqrt{6}} = 0.1304 \frac{mg}{kg} \approx 0.13 \frac{mg}{kg}$$

Alternatively, students can use Excel's function VAR () and STEDV () to quickly calculate variance and standard deviation, respectively (see Chapter 3, Table 3.2).

Chapter 2

Questions

1) **Which of the following is preferred in reporting arsenic concentration in drinking water: 0.04 mg/kg, 40 µg/L, 40 ppb, 40,000 ng/L?**
   The mass/volume unit of 40 µg/L is preferred in reporting, although its equivalent 40 ppb is also frequently seen. The unit of mg/kg is not correct in reporting chemicals in water. Besides, a large number such as 40,000 is customarily reduced using a large mass unit.

2) **If one measures the water sample of 95.2 mL using a graduated cylinder and then measures the same water sample of 5.63 mL using a pipette, the combined volume with the correct significant figure in reporting is: 100.83 mL, 100.8 mL, 101 mL, or 100 mL?**
   The correct answer is 100.8 mL. For addition and subtraction of two or more numbers, the answer should be rounded to the same number of decimal places as the measurement with the least number of decimal places.

3) **If one dilutes a working solution of COD standard containing 125 mg/L potassium acid phthalate by pipetting 5.95 mL into a volumetric flask to make a final volume of 50.00 mL, the final concentration reported in the correct significant figure is: 14.875 mg/L, 14.88 mg/L, 14.9 mg/L, 15 mg/L?**
   The correct answer is 14.9 mg/L, since both 125 and 5.95 have three significant figures. For multiplication and division of two or more numbers, the product or quotient will have the minimum number of significant figures.

4) **Explain why in the National Ambient Air Quality Standards, the concentration units of $PM_{2.5}$, PM_{10}, and lead are expressed in $µg/m^3$, and the concentration units of other pollutants (CO, NO_2, SO_2, O_3) are ppm or ppb? Can you convert 10 $µg/m^3$ of $PM_{2.5}$ into ppm? Can you convert 10 ppm SO_2 into $µg/m^3$?**
   The concentrations of atmospheric chemicals in their solid form such as Pb and PM are reported only in mass per volume unit such as $µg/m^3$. These cannot be converted into the volume/volume units such as ppm_v. For gaseous compounds (CO, NO_2, SO_2, O_3), both mass/volume and volume/volume can be reported; thus, 10 ppm_v SO_2 can be converted into $µg/m^3$ (see Eq. 2.5).

5) **Which one of the following defines 1 M of HCl: (a) 1 mol, (b) 1 mol/L, (c) 1 mol/kg?**
   1 mol/L is the same as 1 M.

6) **Which one of the following defines 1 N of NaOH: (a) eq/L, (b) meq/L, (c) mol/L, (d) 1 mol/kg?**
   1 eq/L is the same as 1 N.

Fundamentals of Environmental Sampling and Analysis, Second Edition. Chunlong Zhang.
© 2024 John Wiley & Sons, Inc. Published 2024 by John Wiley & Sons, Inc.
Companion Website: www.wiley.com/go/EnvironmentalSamplingandAnalysis2e

7) **Can an analytical method have good precision without the desired good accuracy, and vice versa?**

Good precision will not guarantee good accuracy. For example, the concentrations from multiple measurements of a sample can be very close (good precision), but the average concentration could be far from its actual concentration (poor accuracy).

Good accuracy usually but not always leads to good precision. For example, the concentrations from multiple measurements could be very dispersed (poor precision), but coincidentally (the chance may be small) the average concentration from these multiple measurements could be identical to the actual concentration of the sample (good accuracy).

8) **What is the ideal value of % recovery? Can an analytical procedure have a recovery exceeding 100%?**

An ideal analytical recovery is 100%. In the real world, recoveries of less than 100% are more commonly seen, but recoveries of higher than 100% are also possible particularly when positive interferences exist during analysis.

9) **Explain the difference between standard deviation and relative standard deviation (RSD), and why RSD is better to represent precision?**

RSD is expressed in percentages and is obtained by multiplying the standard deviation (s) by 100 and dividing this product by the average (Eq. 1.6). Unlike s that has the unit (e.g., concentration), RSD is normalized by the average. Thus, the value of RSD in % better relates to the precision. For example, a 5% RSD is generally considered to be precise, whereas 40% is not. On the contrary, the s value itself does not indicate the order of magnitude of the precision. For example, a measurement of 500 ± 1.0 (RSD = 0.2%) with a larger s (1.0) is actually more precise than the measurement of 5.00 ± 0.5 (RSD = 10%) with a smaller s (0.5).

10) **What is the purpose of a standard calibration curve?**

A standard calibration curve is obtained by plotting the signal of the measurements vs. the concentrations of the standards with known concentrations. It is then used to calculate the unknown concentrations in the samples according to the measured signals of the samples.

11) **What are the common sources of errors in preparing a calibration curve?**

A calibration curve is a plot of analytical signal (e.g., absorbance, in absorption spectrophotometry) vs. concentration of the standard solutions. Therefore, the main sources of error are the errors associated with the standard concentrations and the measured signals. Concentration errors depend mainly on the accuracy of the weighting device (balance) and volumetric glassware (volumetric flasks, pipettes, solution delivery devices) and on the precision of their use by the persons preparing the solutions. Signal measurement error depends largely on the instrumental method used and on the concentration of the analyte. Instrumental errors can be reduced through repeat measurements to average the random noise (particularly important at lower concentrations), or through the change in instrumental conditions (such as the slit width or the light path length) to improve the signal-to-noise ratio.

12) **Indicate the category of the following organic compounds from this list: (1) aliphatic acid, (2) aliphatic ether, (3) saturated aliphatic hydrocarbon, (4) polycyclic aromatic hydrocarbon, (5) aliphatic alcohol, (6) aliphatic amine, (7) unsaturated aliphatic hydrocarbon, (8) chlorinated hydrocarbon.**

(a) $CH_3CH_2NH_2$

(b) CH_3CH_2COOH

(c) $CH_3CH_2CH_2OH$

(d) $CH_3CH_2OCH_3$

(e) Cl, Cl, Cl (chlorinated alkene structure)

(f) CH_2, CH_2 (diene structure)

(g) H_3C, CH_3

(h) (naphthalene structure)

a) aliphatic amine
b) aliphatic acid
c) aliphatic alcohol
d) aliphatic ether
e) chlorinated hydrocarbon
f) unsaturated aliphatic hydrocarbon
g) saturated aliphatic hydrocarbon
h) polycyclic aromatic hydrocarbon

13) **Name the following functional groups: (a) -COOCH$_3$, (b) -CONH$_2$, (c) CHO, (d) -COOH.**
(a) ester, (b) amide, (c) aldehyde, (d) carboxylic acid

14) **Name the following functional groups: (a) -OH, (b) -NO, (c) C≡N, (d) -R-N-N=O.**
(a) alcohol, (b) nitro, (c) nitrile, (d) nitrosoamine

15) **Name the following compounds:**

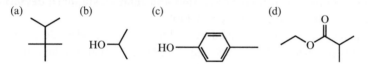

(a) 2-methyl-3,3-dimethyl butane, (b) 2-propanol, (c) para methyl phenol, (d) methyl ethyl propionate.

16) **Name the following compounds:**

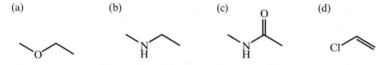

(a) methyl ethyl ether, (b) methylethylamine, (c) methylacetamide, (d) chloroethane.

17) **Name the following compounds:**

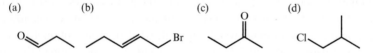

(a) propanal, (b) 5-bromo-3-pentene, (c) 2-butanone, (d) 3-chloro-2-methylpropane.

18) **Draw the structure of the following compounds: (a) 2,3-dichloropentane, (b) butyric acid, (c) 2-methyl-2-butene, (d) 1-bromobutane.**

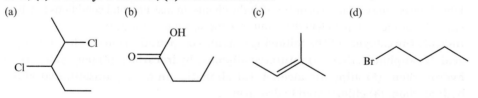

19) **Draw the structure of the following compounds: (a) trimethylamine, (b) ethyl alcohol, (c) methylpropylacetylene, (d) *o*-dimethylbezene.**

(a)   (b)   (c)   (d)

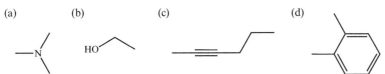

20) **Draw the structure of the following compounds: (a) ethyl methyl ether, (b) 2,2,3-trimethylbutane, (c) 2-butanol, (d) 2-bromo-3-methyl butane.**

(a)   (b)   (c)   (d)

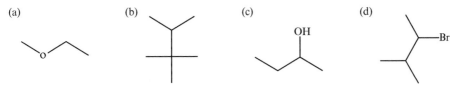

21) **Draw the structure of the following compounds: (a) 2-methyl-3-pentanone, (b) benzoic acid, (c) phenyl acetate, (d) methyl acetate.**

(a)   (b)   (c)   (d)

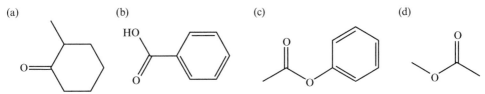

22) **Identify all the functional groups in each of the following environmental pollutants: a) testosterone, (b) *N*-nitrosodimethylamine, (c) glyphosate, (d) uric**

(a)   (b)   (c)   (d)

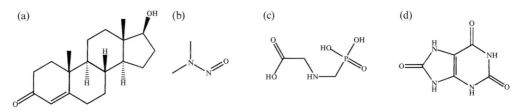

(a) ketone, phenol, carbonyl, (b) nitrosoamine, (c) carboxylic acid, carbonyl, phosphonic acid, (d) ketone, pyrrole, carbonyl.

23) **Identify all the functional groups in each of the following environmental pollutants: (a) 2,4-D, (b) (5-hydroxynaphthalen-1-yl) *N*-methylcarbamate, (c) 2-amino-2-methyl propanol, (d) 2-methyl propanal.**

(a)   (b)   (c)   (d)

(a) carboxylic acid, carbonyl, ether, aryl halide (chlorobenzene), (b) amine, carbonyl, ether, phenol, (c) amine, hydroxyl, (d) hydroxyl.

24) **Identify all the functional groups in each of the following environmental pollutants: (a) p-cresol, (b) di-ethylphthalate, (c) 1,3-butanediol, (d) cyclohexyl isocyanate.**

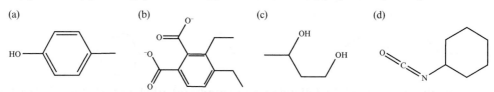

(a)   (b)   (c)   (d)

(a) phenol, (b) carboxyl, carbonyl, (c) alcohol, (d) carbonyl (C=O), imine (C=N).

25) **Identify all the functional groups in each of the following environmental pollutants: (a) 1, 2, 3-thiadiazole, (b) 2-hydrazinopyridine, (c) 2,2,4 trimethyl pentane, (d) 2-octanone.**

(a)   (b)   (c)   (d)

a) Thiadiazole is a five-membered heterocyclic system containing two nitrogen atoms and one sulfur atom.

b) Pyridine is a basic heterocyclic organic compound structurally related to benzene, with one methine group (=CH−) replaced by a nitrogen atom. -NH-NH$_2$ is called hydrazino group.

c) It is made of an alkane with three methyl groups.

d) It is an 8-carbon ketone with one carbonyl group (C=O).

26) **Name three structurally related isomers that could be related to the commercial use of DDT due to its impurities and degradation in the environment.**

Commercial DDT is a mixture of the desired *p,p'*-DDT (77%) and impurity *o,p'*-DDT (15%). *p,p'*-DDE (dichlorodiphenyldichloroethyene) and *p,p'*-DDD (dichlorodiphenyldichloroethane) make up the balance of impurities and are also the degradation products in the environment.

(a) *p, p'*-DDT

(b) *o, p*-DDT

(c) *p, p'*-DDE

(d) *p, p'*-DDD

27) **Draw the structures of (a) 1,2,3,7,8-pentachlorodibenzo-*p*-dioxin and (b) 2,3,4,7,8-pentachlorodibenzofuran.**

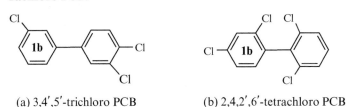

(a) 1,2,3,7,8-pentachlorodibenzo-p-dioxin   (b) 2,3,4,7,8-pentachlorodibenzofuran

28) **Draw the structure of two PCB congeners: (a) 3,4′,5′-trichloro PCB and (b) 2,4,2′,6′-tetrachloro PCB.**

(a) 3,4′,5′-trichloro PCB   (b) 2,4,2′,6′-tetrachloro PCB

29) **Which of the following PCB congener exhibits the coplanar structure and hence the highest toxicity: (a) 2,3,2′,6′ PCB, (b) 2,4,4′,5′ PCB, (c), 3,4,5,4′ PCB, (d) 3,4,3′,4′PCB?**
The coplanar (non-ortho-substituted) congeners of PCBs have a fairly rigid structure with their two phenyl rings in the same plane, presenting more toxicity similar to dioxins. "Non-ortho" means Cl-substitutions at the 3, 4, 5, 3′, 4′, 5′ positions only, excluding 2, 6, 2′, 6′ positions. Thus, (c) and (d) are coplanar PCBs with higher toxicity.

30) **Give the number of congeners for the following halogenated aromatic pollutants: (a) PCBs, (b) PCDDs, (c) PCDFs, (d) PBDEs.**
(a) 209, (b) 75, (c) 135, (d) 209

31) **What structural feature makes halogenated organic pollutants chemically stable and persistent in the environment?**
The very strong C–X bond (particularly C–F bond) in halogenated hydrocarbons makes them chemically stable and environmentally persistent.

32) **What makes organic metallic compounds more of environmental concern than their free metal ions?**
Organometallic compounds have the unique bond between carbon and metals (Hg, Pb, Sn). Such bonds make most organometallic compounds highly mobile, persistent, easily bioaccumulated, and highly toxic. Compared to their inorganic elemental forms or cationic metal ions, the organometallic compounds are often more toxic than their inorganic species.

33) **Do congeners with the same number of Cl or Br have the same toxicity?**
Toxicity could vary considerably depending on the number and locations of the Cl and Br in the halogenated congeners.

34) **What were the main uses of: (a) PCE, (b) PCBs, (c) PBDEs, (d) PCP?**
(a) Solvents for degreasers in electroplating and commercial dry cleaning, (b) dielectric fluid in transformers until the ban in the 1970s, (c) fire retardants in furniture foam, plastics for TV cabinets, consumer electronics wire insulation, (d) fungicides, preservative, and disinfectants.

Problems

1) **A 101.5 g of moist soil sample is dried at 105 °C in an oven overnight, and the dried soil sample weighs 90.4 g. What is the % moisture content in soil reported on dry basis, and wet basis, respectively?**

% MC (dry basis) $= (101.5 - 90.4)/90.4 = 12.3\%$

% MC (wet basis) $= (101.5 - 90.4)/101.5 = 10.9\%$

2) **The *p,p'*-DDT residue in a grape sample was reported to be 78 μg/kg (dry basis). The fresh grapes contain 81% moisture (wet basis). What is the concentration of *p,p'*-DDT reported on wet basis?**

From Eq. 2.4: $C_d = C_w/(1 - \% \text{ MC})$

$C_w = C_d \times (1 - \% \text{ MC}) = 78 \times (1 - 0.81) = 14.82$ μg/kg

3) **The moisture content of dewatered sludge from a belt press unit is 19% (wet basis). The residual chromium (Cr) is measured to be 245 mg/kg (wet basis). What is the Cr concentration on dry basis?**

Apply Eq. 2.4: $C_d = C_w/(1 - \% \text{ MC}) = 245/(1 - 0.19) = 302$ mg/kg.

4) **A water sample has 10 mg/L Hg^{2+} and a density of 1.0 g/mL (atomic weight of Hg = 200.59). Calculate the following: (a) Hg^{2+} in ppb, (b) Hg^{2+} in μM, and (c) the number of Hg^{2+} ions in 1 L of this sample containing 10 mg/L Hg^{2+} (Avogadro's number = 6.022 × 10^{23} /mol).**

a) 1 μg/L $=$ 1 ppb. So, 10 μg/L $=$ 10 ppb.

b) $10 \dfrac{\mu g}{L} \times \dfrac{1\,\mu\text{mol}}{200.59\,\mu g} = 4.99 \times 10^{-2}$ μM

c) $4.99 \times 10^{-2} \dfrac{\mu\text{mol}}{L} \times \dfrac{1\text{ mol}}{10^{6}\,\mu\text{mol}} \times \dfrac{6.022 \times 10^{23}}{1\text{ mol}} = 3.00 \times 10^{16}$ molecules per liter

5) **The concentrations of arsenic (As) and selenium (Se) in a drinking water well were 2.0 and 3.8 ppb. (a) Convert arsenic concentration into ppm and mg/L. (b) Convert selenium concentration into molarity (M) and micromolarity (μM). (c) Have the concentrations exceeded the maximum contaminant level (MCL) of 50 μg/L for both elements? The atomic weight of Se is 79.**

a) 2.0 ppb $=$ 2.0 μg/L $=$ 0.002 ppm $=$ 0.002 mg/L

b) $3.8 \text{ ppb} = 3.8 \dfrac{\mu g}{L} \times \dfrac{1\,\mu\text{mol}}{79\,\mu g} = 0.048 \dfrac{\mu\text{mol}}{L} = 0.048 \text{ μM} = 4.8 \times 10^{-8}$ M

c) They have not exceeded the MCL of 50 mg/L, because 2.0 μg/L and 3.8 μg/L are less than 50 μg/L.

6) **Carbon monoxide (CO) and hydrocarbons (HC) are the two main exhaust gases produced by the combustion of gasoline-powered vehicles. Their concentrations are regulated by required car inspections in some states. (a) Convert the regulatory standard of 220 ppm HC into mg/m³ (assuming a nominal molecular weight MW = 16). (b) If the actual exhaust concentration of CO (MW = 28) at a low-speed emission test is 0.16%, what is the CO concentration in ppm and mg/m³?**

a) $220 \text{ ppm} \times \dfrac{MW}{24.5} = 220 \text{ ppm} \times \dfrac{16}{24.5} = 144 \dfrac{mg}{m^3} = 1.44 \times 10^{5} \dfrac{\mu g}{m^3}$

b) $0.16\% = \dfrac{0.16}{100} = \dfrac{0.16 \times 10^{4}}{10^{6}} = \dfrac{1600}{10^{6}} = 1600 \text{ ppm}$ $1600 \text{ ppm} \times \dfrac{MW}{24.5} = 1600 \text{ ppm} \times \dfrac{28}{24.5} = 1829 \dfrac{mg}{m^3}$

7) **The exhaust gas from an automobile contains 0.002% by volume of nitrogen dioxide (MW = 46). What is the concentration of NO$_2$ in mg/m^3 at standard temperature and atmosphere pressure (25°C and 1 atm pressure)?**

$$0.002\% = \frac{0.002}{100} = \frac{0.002 \times 10^4}{10^6} = \frac{20}{10^6} = 20 \text{ ppm}$$

$$20 \text{ ppm} \times \frac{MW}{24.5} = 20 \text{ ppm} \times \frac{46}{24.5} = 37.6 \frac{mg}{m^3} = 3.76 \times 10^4 \frac{\mu g}{m^3}$$

8) **The 8-h ozone (O$_3$) concentration is 175 μg/m^3 on a summer day (35 °C and 1 atm). Does this exceed the 8-h standard of 0.075 ppm$_v$?**
Apply Eq 2.6: Conversion factor $V = nRT / p = 1 \times 0.082 \times (273 + 35)/1 = 25.256$
MW of O$_3$ = 48,
Apply Eq. 2.5: C = $175 \times 25.256/48 = 92.1$ ppb = 0.092 ppm.
Since 0.092 > 0.075 ppm$_v$, it exceeds the 8-h standard.

9) **An analytical method is being developed to measure trichloracetinitrile in water. The analyst has performed a number of spiked and recovery experiments in different water matrices to determine the method detection limits (MDLs). The data are reported in the following table.**

Trichloroacetonitrile recovery and precision data					
Matrix	Number of sample (n)	True conc. (μg/L)	Mean conc., $\bar{x}$ (μg/L)	Standard deviation (s)	Relative standard deviation (% RSD)
Distilled Water	7	4.0	2.8	0.190	6.8
River Water	7	4.0	4.55	0.314	6.9
Ground Water	7	4.0	5.58	0.145	2.6

a) **Which water matrix yields the most precise results? Why?**
b) **Which water matrix yields the most accurate results? Why?**
c) **Calculate the MDLs with a 98% confidence level for each water matrix.**

    a) The analysis for ground water matrix is the most precise, because it has the lowest % RSD (2.6%) of all.

    b) The analysis for river water matrix is the most accurate, because the measured concentration (4.55 μg/L) is the closet to the true value (4.0 μg/L). Alternatively, we can use the % recovery to rank them, % recovery = analytical value/true value × 100:

        Distilled water: 2.8/4.0 × 100 = 70%
        River water: 4.55/4.0 × 100 = 114%
        Groundwater: 5.58/4.0 × 100 = 140%

    Since the recovery of river water analysis is closest to 100%, it is the most accurate analysis.

    c) MDL = $s \times t$, for $n = 7$, $df = 7 - 1 = 6$, $t = 3.143$ at 99% (one-sided) confidence level (Appendix D2).

        Distilled water: 0.190 × 3.143 = 0.597 μg/L
        River water: 0.314 × 3.143 = 0.987 μg/L
        Groundwater: 0.145 × 3.143 = 0.456 μg/L

    Hence, the method for groundwater matrix has the lowest method detection limit.

10) **The relative percentage differences (RPD) were determined for the comparison of six analytical methods cited in Table 2.2 for both source water and treated water (Batt et al., 2017). Comment on the analytical performance based on the median RPD given below (n.a. = not available).**

Method number	Water type	Number of replicate pairs	Median RPD (%)
Method 1	Source water	52	14.3
	Treated water	12	9.4
Method 2	Source water	43	11.2
	Treated water	6	5.6
Method 3	Source water	35	8.7
	Treated water	20	6.4
Method 4	Source water	12	13.1
	Treated water	5	9.1
Method 5	Source water	16	7.6
	Treated water	22	18.4
Method 6	Source water	4	15.8
	Treated water	0	n.a.

RPD values represent the precision based on two measurements. The small RPD obtained indicates the acceptable precision of all six analytical methods regardless of the two different water matrices. For Methods 1 to 4, the precision for the measurement of source water appears to be better than that of the treated water.

11) **Method 4 in Batt et al. (2017) developed by Schultz and Furlong (2008) was for trace analysis of antidepressant pharmaceuticals and their selected degradates in aquatic matrixes by LC/ESI/MS/MS. Batt et al. (2017) suggested eight out of the 11 analytes listed below be excluded for quality purpose. Provide the justifications based solely on the overall recovery (%) data in three types of water samples.**

Analyte	Distilled water spikes	Source water spike	Treated water spike
Citalopram	87	37	52
Duloxetine	21	10	32
Fluoxetine	54	34	45
Fluvoxamine	41	25	38
Norfluoxetine	53	27	28
Norsertraline	48	36	59
Paroxetine	21	23	35
Sertraline	41	31	36

The analytical recoveries of Method 4 for these eight compounds were mostly low for all three sample matrices. This method needs to be further optimized for valid uses.

12) **A Cr^{6+} solution of 0.175 mg/L was prepared and analyzed eight times over the course of several days. The results were 0.195, 0.167, 0.178, 0.151, 0.176, 0.155, 0.154, and 0.164 mg/L. The standard deviation of this data set was calculated to be 0.0149 mg/L. (a) Calculate the method detection limit (MDL). (b) What is the estimated practical quantitation limit (PQL)? (c) Is it accurate to measure samples containing Cr^{6+} around 0.1 mg/L?**

    a) $s = 0.0149$, $n = 8$, $df = 7$, $t = 2.998$ at 99% confidence level (one-sided), so: MDL = 0.0149 × 2.998 = 0.0447 mg/L.

    b) If we use PQL of 2 – 10 times of MDLs, then PQL = 0.0893 – 0.447 mg/L.

    c) Since PQL (0.0893 – 0.447 mg/L) is dependent on sample matrix, the above method would be ok for clean sample matrix like groundwater. This, however, may not be accurate enough if we have a complex matrix.

13) **The following absorbance data at wavelength 543 nm were obtained for a series of standard solutions containing nitrite (NO_2^-) using a colorimetric method: $A = 0$ (blank), $A = 0.220$ (5 μM), $A = 0.41$ (10 μM), $A = 0.59$ (15 μM), $A = 0.80$ (20 μM). Using Excel (a) to plot the calibration curve, and (b) to determine the calibration equation and the coefficient of determination (R^2)**

    a) The calibration plot is shown below.

    b) The calibration equation from Excel is also shown in the graph: y (absorbance) = 0.0378 x (μM) + 0.01 ($R^2 = 0.9817$).

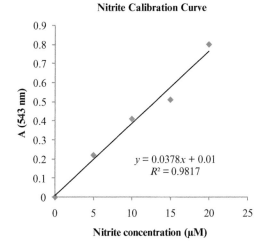

14) **Pyrene was analyzed using GC-MS at the selected ion mode. The calibration curve of peak area (GC-MS response) vs. concentration is shown below along with the regression output. A blank sample was analyzed seven times, giving a mean peak area of 125 and standard deviation of 15. A check standard at 0.5 ppm was also measured seven times, giving a standard deviation of 300. The raw data, Excel's printouts of linear regression along the line plot are attached. (a) Perform your own Excel analysis and compare it with what is given: $y = 72024.7 x + 107.2$. (b) What is the lin-**

ear regression equation and the R^2? (c) What is the calibration sensitivity? (d) Is the linear regression statistically significant, if yes then why? (e) If an unknown liquid sample (after a dilution of 10 times) was injected into GC-MS and gave a peak area of 35720, what is the concentration of this sample?

Concentration of pyrene (ppm)	GC-MS response
0	0
0.1	7456
0.2	14929
0.3	21219
0.4	28783
0.6	43496

Summary output

Regression statistics

Multiple R	0.999793298
R^2	0.999586638
Adjusted R Square	0.999483298
Standard error	353.7486093
Observations	6

	SS	MS	F	Significance F
Regression	1210430543	1210430543	9672.75953	6.408E-08
Residual	500552.3143	125138.0786		
Total	1210931095			

	Coefficients	Standard error	t Stat	p-Value	Lower 95%	Upper 95%	Lower 95.0%	Upper 95.0%
Intercept	107.2428	242.8861	0.44153	0.68164403	-567.11851	781.60	-567.12	781.60
Concentration	72024.71	732.3291	98.3501	6.4084E-08	69991.438	74057.99	69991.4	74057.99

a) The Excel's linear regression output along with a calibration curve should match what are given: $y = 72024.7\,x + 107.2$.
b) The slope and intercept can be read from the printed output: slope = 72024.7, intercept = 107.2; hence, the linear regression is: $y = 72024.7\,x + 107.2$, where x = pyrene concentration in ppm, and y = GC-MS response. $R^2 = 0.99948$.
c) Calibration sensitivity = Slope = 72024.7 per unit ppm.
d) F (9672.7) > $F_{critical}$ (6.408E-08), so it is significant. This can also be told by the p-value of the slope ($p = 6.4084E-08 < 0.05$).
e) $y = 35720$, $x = (35720 - 107.2)/72024.7) = 0.494$ ppm. Because a dilution factor of 10 was used prior to GC-MS analysis, the original sample has a concentration of $10 \times 0.494 = 4.94$ ppm.

Chapter 3

Questions

1) **Describe in what cases the use of "median" can be preferred over that of "mean" in environmental data reporting.**

Arithmetic "mean" is commonly used in environmental reporting. However, the use of "median" can be preferred: (a) If there are some extreme values or data distributions are skewed due to the presence of outliers, (b) when nonnumerical data exist such as nondetect, or below detection limit. "Media" is less affected by extreme data points, and unlike "mean," it is not affected by data transformations.

2) **Evaluate the pros and cons when using mean or median to evaluate the following two data sets (unit in mg/L) – Data A: 1, 1, 1,1, ...1, 10^6 ($n = 1,000$). Data point 10^6 could be an outlier. Data B: 0.11, ND, 0.13, 0.15. ND denotes "not detected."**

For the first data set, 10^6 mg/L is apparently an outlier. This may be simply a typo during the data input process. It should be removed. The mean for all data points is $(1 \times 999 + 10^6)/1000 = 1000.999$ mg/L, and the median is 1 mg/L, which is not affected by the outlier of 10^6.

For the second data set, a "not detected" data is a nonnumerical data (called the "censored data"). The mean and standard deviation of such data cannot be calculated. Hence, it cannot be compared with a legal standard. Instead, a numerical value of 0.13 µg/L in the form of median can be reported.

3) **For what type of data distribution is the use of geometric mean and geometric standard deviation preferred over the use of arithmetic mean and arithmetic standard deviation?**

For log-normal distribution, geometric mean ($\overline{x}_g$) and geometric standard deviation (s_g) are preferred. The usual arithmetic mean ($\overline{x}$) and arithmetic standard deviation (s) are the correct choice only for normally distributed data.

4) **Explain the difference between one-way ANOVA and the F-test.**

The F-test is used to test the variance, while one-way ANOVA tests the mean. The F-test uses the ratio of the mean-squared error between the two groups, whereas ANOVA separates the within-group variance from the between-group variance.

5) **Define Type I error and Type II error. Explain why both "false positive" and "false negative" should be avoided in the analysis and monitoring of environmental contaminants.**

By definition, Type I error is to reject a null hypothesis when it is true. Type II error is to accept a null hypothesis when it is false. Equivalent to a more understandable justice case, Type I is to put innocent people in a jail, whereas Type II error is like setting criminals free. In the environmental analysis field, Type I error is for someone to report an area that is contaminated

Fundamentals of Environmental Sampling and Analysis, Second Edition. Chunlong Zhang.
© 2024 John Wiley & Sons, Inc. Published 2024 by John Wiley & Sons, Inc.
Companion Website: www.wiley.com/go/EnvironmentalSamplingandAnalysis2e

when it is actually not (which costs unnecessary resources to clean the site), and Type II error is to falsely report that an area is clean when it is actually contaminated (in this case, health and environmental risk may result because the analyst did not detect any contaminants). Both errors should be avoided. However, minimizing one type of error will increase another type of error.

6) **Define "outliers" and also describe how the "outlier" data should generally be dealt with and be removed.**

Outliers are observations that appear to be inconsistent with the remainder of the data collected. The basic philosophy in dealing with outlines is not to discard the outliers without a reason. Remedies are: (a) to replace the data points if redoing sampling and analysis are possible and economically feasible, (b) to remove the outlier if a sound statistics can be based, or (3) to retain the outliers and use more robust statistical methods that are not seriously affected by a few outliers. Keep in mind that "outliers" may not be actually the "outliers." In environmental monitoring, they may present the true inherent temporal/spatial variation of data or samples from "hot spot," for example from accidental spills.

7) **Can a data point be excluded solely based on one of the outlier tests? Why or why not?**

No data should be discarded solely based on one of the statistical tests. One should first examine the cause of possible outlier, if at all possible. The outlier may be the true outlier because of mistakes such as sampling error, analytical error (instrumental breakdowns, calibration problems), transcription, keypunch, or data-coding error. The outlier may be from an inherent spatial/temporal variation of data or unexpected factors of practical importance such as malfunctioning pollutant effluent controls, spills, plant shutdown, or hot spots. In these cases, the suspected outlier is actually not the true outlier.

8) **Why is it not a good practice to either delete the nondetected data or substitute the nondetected data?**

Deletion of nondetected censored data is probably the worst procedure and should never be used because it causes a large and variable bias in the parameter estimates. The use of substitution method, i.e., to substitute the "less than" values with zero, half of the LOQ, or LOQ, is another common mistake. This substitution method produces estimates of the mean and median that are biased low (substitution with zero) or biased high (e.g., substitution with LOQ or detection limit).

9) **Grubbs, Dixon, and Rosner are three commonly used methods for the detection of outlying data points. Describe the preferred usage of each method in terms of the number of outlier(s) to be detected and number of data points (n).**

The Grubbs test should not be used for sample sizes of six or fewer since it frequently tags most of the points as outliers. Dixon's test is used for a small number of outliers with a sample size between 3 and 30. Dixon's test is vulnerable to "masking," in which the presence of several outliers masks the detection of even one outlier. Hence, Rosner's test is recommended by the USEPA for detecting multiple outliers of large data sets ($n \geq 20$).

10) **Describe situations and types of data when nonparametric statistics should be used.**

Nonparametric tests can be applied to situations when the data does not follow any probability distribution, when the data constitutes of ordinal values or ranks, and when there are outliers in the data.

11) **What test is the nonparametric alternative to the parametric: (a) one-way ANOVA and (b) two-sample t-test?**

(a) Kruskal Willis test is the nonparametric alternative to the one-way ANOVA. (b) Mann Whitney test is the nonparametric alternative to the two-sample t-test.

12) **R literally can do everything in data analysis. What makes R a powerful data analysis programming language? Give a list of statistical methods that R can handle for environmental data, while Excel cannot.**

It is because of the large number of R packages available for users to install and use. R packages include code, data, and documentation in a standardized collection format that can be easily installed by users of R. For example, Excel doesn't have the ability to do statistical tests of nonnormal (i.e., not "bell-shaped") data. In contrast, the distribution-free nonparametric statistical packages are readily available in R. The R package on the statistics for censored environmental data, developed by Dennis Helsel, is another example.

13) **What package has been developed to do specifically the censored nondetect environmental data?**

NADA and NADA2. NADA stands for nondetects and data analysis. The description of R packages for NADA2 (version 1.1.3) is available: https://cran.r-project.org/web/packages/NADA2/NADA2.pdf

Problems

1) **Calculate (a) arithmetic mean and standard deviation, (b) geometric mean and standard deviation of 15 TSP data points in Example 3.5 (2.56, 3.33, 3.43, 3.64, 3.69, 3.74, 3.76, 3.76, 3.97, 4.04, 4.08, 4.29, 4.33, 4.34, 4.47). Use Excel's functions.**

a) The Excel functions for arithmetic mean and standard deviation are AVERAGE( ) and STDEV( ), respectively. Thus,
Arithmetic mean = AVERAGE ( ) = 3.828667
Arithmetic standard deviation = STDEV ( ) = 0.489458

b) The Excel functions for geometric mean and geometric standard deviation are GEOMEAN( ) and EXP(STEDV(LN( ))), respectively. Thus,
Geometric mean = GEOMEAN( ) = 3.79602
Geometric standard deviation = EXP(STDEV(LN( ))) = 1.150046

Data		Excel Function		
2.56	(a)			
3.33		Arithmetic mean:	AVERAGE( )	3.828667
3.43		Standard deviation:	STDEV( )	0.489458
3.64				
3.69	(b)			
3.74		Geometric mean:	GEOMEAN( )	3.79602
3.76		Standard deviation:	EXP(STDEV(LN( )))	1.150046
3.76				
3.97				
4.04				
4.08				
4.29				
4.33				
4.34				
4.47				

2) **The aluminum concentrations (g/kg) in Elrama School Superfund site in Washington County, PA were reported (Sing et al., 1997): 31.9, 8.03, 12.2, 11.3, 4.77, 5.73, 5.41, 8.42, 8.20, 9.01, 8.60, 9.49, 9.53, 7.46, 7.70, 13.7, 30.1, 7.03, 2.73, 5.82, 8.78, 0.36, 7.05 ($n = 23$). (a) Conduct the summary descriptive statistics analysis. (b) Use Excel's functions to confirm the arithmetic mean, arithmetic standard deviation, median, and variance estimated from the descriptive statistics.**

	Al (g/kg)	(a) Descriptive statistics		(b) Use of Excel Functions	
1	31.9				
2	8.03				
3	12.2	Mean	9.709565	AVERAGE( )	9.709565
4	11.3	Standard Error	1.524244		
5	4.77	Median	8.2	MEDIAN( )	8.2
6	5.73	Mode	#N/A		
7	5.41	Standard Deviation	7.31002	STDEV( )	7.31002
8	8.42	Sample Variance	53.43639	VAR( )	53.43639
9	8.20	Kurtosis	5.48113		
10	9.01	Skewness	2.286158		
11	8.60	Range	31.54		
12	9.49	Minimum	0.36		
13	9.53	Maximum	31.9		
14	7.46	Sum	223.32		
15	7.70	Count	23		
16	13.7	Confidence level (95.0%)	3.16109		
17	30.1				
18	7.03				
19	2.73				
20	5.82				
21	8.78				
22	0.36				
23	7.05				

3) **Use TINV(α, df) to calculate the t-value at 95% confidence interval (two-sided) for the data in Problem 2. Use CONFIDECE.T (α, s, n) to confirm the confidence interval obtained from the output of descriptive statistics in Problem 2.**

Using Excel function TINV(α, df) for $\alpha = 0.05$ (95% confidence) and $df = n-1 = 23-1 = 22$, TINV (0.05, 22) = 2.073873.

Using Excel function CONFIDENCE(α, STDEV( ), n) will result in 3.16109, where $\alpha = 0.05$ for 95% confidence, STDEV( ) $= s = 7.31002$, and $n = 23$. The value 3.16109 obtained from Excel function is the same as the output of descriptive statistics (see Problem 2). Alternatively, the t-value calculated using TINV(α, df) can be plugged into Eq. 3.9 to calculate the confidence interval CI:

$$CI = \bar{x} \pm t_{n-1,1-\frac{\alpha}{2}}\left(\frac{s}{\sqrt{n}}\right) = 9.709565 \pm 2.073873\left(\frac{7.31002}{\sqrt{23}}\right)$$

$$= 9.709565 \pm 3.16109, \text{ or } 9.71 \pm 3.16 \text{ g/kg for reporting purpose.}$$

4) **For the data in Problem 2, use Excel's function to calculate (a) the geometric mean and geometric standard deviation, (b) Q_1, Q_3, and *IQR*.**

   a) The Excel functions for geometric mean and geometric standard deviation are GEOMEAN( ) and EXP(STEDV(LN( ))), respectively. Thus,

      Geometric mean = GEOMEAN( ) = 7.5351

      Geometric standard deviation = EXP(STDEV(LN( ))) = 2.3280

   b) The Excel function for quartile is: QUARTILE(array, quart), where array is the array of data, and quart is the quartile one would like to calculate (quart =1 for Q_1, and quart = 3 for Q_3. Thus, for the data in Problem 2, we have:

      Q_1 = QUARTILE (( ), 1) = 6.425

      Q_3 = QUARTILE (( ), 3) = 9.51

      $IQR = Q_3 - Q_1 = 9.51 - 6.425 = 3.085$

5) **For the data in Problem 2, does the aluminum concentration data fit better to the log-normal distribution? (a) Construct a histogram of this raw data set. (b) Construct a histogram using the logarithm-transformed raw data (base e).**

The histograms of the raw data and the logarithm-transformed data (natural log) are shown in (a) and (b), respectively. The histograms suggest that the raw data is right-tailed, whereas the transformed data fits better to the bell-shaped normal distribution. This is typical of environmental data.

Al (g/kg)	Al (LN, g/kg)
31.9	3.46
8.03	2.08
12.2	2.50
11.3	2.42
4.77	1.56
5.73	1.75
5.41	1.69
8.42	2.13
8.20	2.10
9.01	2.20
8.60	2.15
9.49	2.25
9.53	2.25
7.46	2.01
7.70	2.04
13.7	2.62
30.1	3.40
7.03	1.95
2.73	1.00
5.82	1.76
8.78	2.17
0.36	−1.02
7.05	1.95

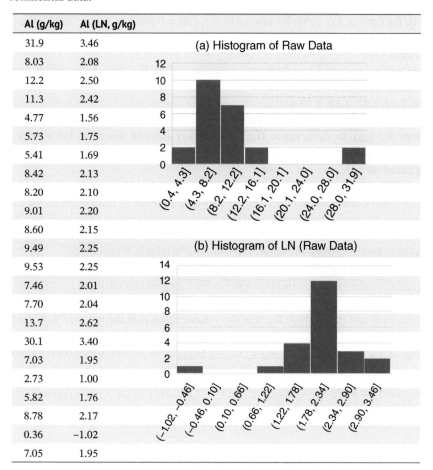

(a) Histogram of Raw Data

(b) Histogram of LN (Raw Data)

6) **The background concentration of lead (Pb) in soils of South Carolina State is normally distributed with a mean of 15.9 mg/kg and a standard deviation of 2.0 mg/kg. Use the z-test to determine if would it be an outlier: (a) if a measurement of 17.5 mg/kg is detected? (b) if a measurement of 40 mg/kg is detected?**

Equation 3.12 is applied to calculate the z value: $z = \dfrac{|x - \bar{x}|}{s}$

a) $x = 17.5$ mg/kg $\rightarrow z = (17.5 - 15.5)/2.0 = 1.0$. Since $z = 1.0 < 3.0$. It is unlikely an outlier according to z-test.

b) $x = 40$ mg/kg $\rightarrow z = (40 - 15.5)/2.0 = 12.25 \gg 3.0$. Since 40 mg/kg is outside the $9.5 - 21.5$ ($\bar{x} \pm 3s : 15.5 \pm 3 \times 2.0$) mg/kg range, the chance is less than 1%. So, it is an outlier, assuming the soil is known not to be contaminated.

7) **Use Table D1 to confirm the probability of 68.27% and 95.44% for the number of data points falling between the range of $\bar{x} \pm 1s$ and $\bar{x} \pm 2s$, respectively, for a normally distributed data set.**

The x range of $\bar{x} \pm 1s$ is equivalent to 0 ± 1.0 in the standard normal distribution, and the x range of $\bar{x} \pm 2s$ is equivalent to 0 ± 2.0 in the standard normal distribution. From Table D1, $P(z < 1) = 0.8413$; thus, $P(0 < z < 1) = 0.8413 - 0.5000 = 0.3413$, the probability for z lies in the range of -1.0 to $+1.0$, $P(-1.0 < z < +1.0)$, is then $0.3413 + 0.3413 = 0.6826$ (68.26% or 68.27% due to rounding-off error).

Similarly, for $z = 0 \pm 2.0$, Table D1 gives $P(0 < z < 2) = 0.9772 - 0.5000 = 0.4772$, the probability for z lies in the range of -2.0 to $+1.0$, $P(-2.0 < z < +2.0)$, is then $0.4772 + 0.4772 = 0.9544$ (95.44%).

8) **Two laboratories analyzed the benzene concentration for the same water sample using the same EPA standard method.**
   Lab 1: 1.13, 1.14, 1.17, 1.19, 1.29 mg/L
   Lab 2: 1.11, 1.12, 1.14, 1.19, 1.25, 1.34 mg/L

   a) **Perform a z-test to determine the most likely outlier for each lab's measurements. What is the underlying assumption of the data distribution that needs to be satisfied in order to use z-test?**

   b) **Perform the Dixon Q-test (i.e., using D_{10}) to determine if an outlier exists.**

      a) The z-test cannot be used; because of the small number of samples here, we cannot assume it is normally distributed. If we know the data is from a normally distributed population, then on inspection it seems that 1.29 (Lab1) and 1.34 (Lab 2) may be the outlier. First calculate the mean and standard deviation for each lab:

      Lab 1: $\bar{x} = 1.184$, $s = 0.0639$; Lab 2: $\bar{x} = 1.192$, $s = 0.0893$.

      $z(1.29) = (1.29 - 1.184)/0.0639 = 1.659 < 3$. So 1.29 is not an outlier.

      $z(1.34) = (1.34 - 1.192)/0.0893 = 1.657 < 3$. So 1.34 is not an outlier.

      b) Dixon's Q-test is used here because the sample size n is in the range of 3–7. Applying the formula in Table 3.7 for D_{10}, we have:

      D_{10} for $1.29 = (x_n - x_{n-1})/(x_n - x_1) = (1.29 - 1.19)/(1.29 - 1.13) = 0.625 < 0.642$ (Table 3.8, $n = 5$, 95%, one-sided). Hence, 1.29 is not an outlier.

      D_{10} for $1.34 = (x_n - x_{n-1})/(x_n - x_1) = (1.34 - 1.25)/(1.34 - 1.11) = 0.391 < 0.560$ (Table 3.8, $n = 6$, 95%, one-sided). Hence, 1.34 is not an outlier.

9) **For the following set of analyses of an urban air sample for carbon monoxide: 325, 320, 334, 331, 280, 331, 338 $\mu g/m^3$, (a) determine the mean, standard deviation, median, mode, and range using Excel's descriptive statistics. (b) Calculate the coefficient of variation and RSD. (c) Is there any value in the above data set that can be discarded as an outlier?**

a) Excel's descriptive statistics:

CO ($\mu g/m^3$)	CO ($\mu g/m^3$)	
325	**Mean**	**322.7143**
320	Standard Error	7.456906
334	**Median**	**331**
331	**Mode**	**331**
280	**Standard deviation**	**19.72912**
331	Sample variance	389.2381
338	Kurtosis	5.100832
	Skewness	−2.18143
	Range	**58**
	Minimum	280
	Maximum	338
	Sum	2259
	Count	7

b) Coefficient of variance (CV) $= s/\bar{x} = \dfrac{19.72912}{322.7143} = 0.0611$ RSD $= CV \times 100 = 6.11\%$

c) First, it appears that 280 is a possible outlier. Using Grubbs' test to calculate the T-value and compare it with the critical value in Table 3.6, $T = \dfrac{\bar{x} - x}{s} = \dfrac{322.7143 - 280}{19.72912} = 2.165 > 1.822$

($T_{critical}$ at 95% confidence level at $df = 7{-}1 = 6$). Hence, 280 is an outlier.

The same calculation can be done for the highest value in the data set, $T = \dfrac{x - \bar{x}}{s} = \dfrac{338 - 322.7143}{19.72912} = 0.775 < 1.822$ ($T_{critical}$ at 95% confidence level at $df = 7{-}1 = 6$). Hence, 338 is not an outlier.

10) **For the raw data set $x_i = 1, 100, 1000$ and its logarithmic (base 10) transformation data set $y_i = 0,2,3$, what are the means and medians for these two data sets? (Hint: For comparison, you need to anti-log the mean and median calculated after log transformation.) What conclusion can be drawn regarding the effect of logarithmic transformation on the values of mean and median?**

The raw data without log transformation has a mean of 367 (average of 1, 100, 1000) and median of 200. For the log-transformed data (0, 2, 3), mean $= (0 + 2 + 3)/3 = 1.667$, median $= 2$. Then we take the anti-log values, mean $= 10^{1.667} = 46.42$, median $= 10^2 = 100$. The log transformation underestimates the actual mean. However, the median is unaffected by the data transformation.

11) **For the data in Problem 2, use Grubbs' and Dixon's tests to determine if the smallest (0.36 g/kg) and the largest (31.9 g/kg) concentrations could be the outliers. The background concentration of aluminum in uncontaminated soil is approximately in the range of 5–30 g/kg (0.5–3%).**

First, we need to sort the raw data from the smallest to the largest: 0.36, 2.73, 4.77, 5.41 ......12.20, 13.70, 30.10, 31.90.

For Grubbs' test, we apply Eq. 3.13a and 3.13b to calculate the T-value:

For 0.36, $T = \dfrac{\bar{x} - x_1}{s} = \dfrac{9.710 - 0.36}{7.31} = 1.279$

For 31.90, $T = \dfrac{x_n - \bar{x}}{s} = \dfrac{31.9 - 9.710}{7.31} = 3.036$

The critical T-value for $n = 23$ can be interpolated from the T-values listed in Table 3.6 as follows using $T (n = 20) = 2.557$ and $T (n = 30) = 3.103$ at 95% confidence:

$T = 2.557 + \dfrac{3.103 - 2.557}{30 - 20} \times (23 - 20) = 2.721$

Since 1.129 < 2.721 and 3.036 > 2.721, the data point of 0.36 is not an outlier but 31.90 is an outlier according to Grubbs' test.

For Dixon's test, $n = 23$, we apply the formula in Table 3.7 to calculate D_{22} for these two suspected outliers, using $x_1 = 0.36$, $x_3 = 4.77$, $x_{n-2} = 13.70$, $x_n = 31.90$.

For 0.36, $D_{22} = \dfrac{x_3 - x_1}{x_{n-2} - x_1} = \dfrac{4.77 - 0.36}{13.70 - 0.36} = 0.331$

For 31.90, $D_{22} = \dfrac{x_n - x_{n-2}}{x_n - x_3} = \dfrac{31.90 - 13.70}{31.90 - 4.77} = 0.671$

The critical D value for $n = 23$ can be interpolated from the D values listed in Table 3.8 as follows:

$T = 0.491 + \dfrac{0.491 - 0.445}{20 - 25}(23 - 20) = 0.463$

Since 0.331 < 0.463 and 0.671 > 0.463, the data point of 0.36 is not an outlier but 31.90 is an outlier. The result from Dixon's test agrees with the result from Grubbs' test.

12) **Using the range of Q_1 - 1.5 × IQR and Q_3 + 1.5 × IQR to determine if the smallest (0.36 g/kg) and the largest (31.9 g/kg) concentrations in Problem 2 could be the outlier.**

From Problem 4, we have calculated: $Q_1 = 6.425$, $Q_3 = 9.51$, and $IQR = 3.085$.

$Q_1 - 1.5 \times IQR = 6.425 - 1.5 \times 3.085 = 1.798$

$Q_3 + 1.5 \times IQR = 9.51 + 1.5 \times 3.085 = 14.138$

Since 31.9 falls out of the range of 1.798–14.138, it is thus an outlier too. The result is consistent with both the Grubbs' test and Dixon's test.

13) **Soil samples were collected at two areas surrounding an abandoned mine and analyzed for lead. At each area several samples were taken. The soil was extracted with an acid, and the extract was analyzed using flame atomic absorption spectrometry. In Area A, Pb concentrations were 1.2, 1.0, 0.9, 1.4 mg/kg. In Area B, Pb concentrations were 0.7, 1.0, 0.5, 0.6, 0.4 mg/kg. (a) Are these two areas significantly different from each other with Pb concentrations at a 90% confidence level? (b) Perform a single-factor ANOVA test (i.e., one-way ANOVA) using Excel.**

The Excel output shows: (a) $F = 10.17818 > F$ critical (5.59146), (b) $p = 0.01527 < 0.1$ (i.e., 90% confidence). Both suggest that the Pb concentrations in Area A and Area B are significantly different from each other at a 90% confidence level.

ANOVA: single factor						

Summary

Groups	Count	Sum	Average	Variance		
Area A	4	4.5	1.125	0.049167		
Area B	5	3.2	0.64	0.053		

ANOVA

Source of variation	SS	df	MS	F	P-value	F crit
Between groups	0.522722	1	0.522722	10.17818	0.01527	5.59146
Within groups	0.3595	7	0.051357			
Total	0.882222	8				

14) **Samples of bird eggs were analyzed for DDT residues. The samples were collected from two different habitats. The data reported are:**

Sample collection area	Number of samples	Mean conc. DDT (ppb)	Standard deviation (s)
Area 1	4	1.2	0.33
Area 2	6	1.8	0.12

a) **Are the two habitats significantly different ($\alpha = 5\%$) from each other in the concentration of DDT to which these birds are exposed?**

b) **Calculate the 90% confidence intervals for DDT concentration in both areas.**

c) **If DDT concentration in Area 2 can be assumed to have a normal distribution with the same mean and standard deviation as the sample measurement listed in the above table (i.e., $\mu = 1.8$ ppb, $\sigma = 0.12$ ppb), what is the probability of a bird egg having residual DDT of greater than 2.2 ppb? less than 2.2 ppb? or between 1.3 and 2.2 ppb?**

a) To determine whether these are significantly different, we will compare $F_{\text{calculated}}$ to F_{critical}. $F_{\text{calculated}} = (0.33)^2/(0.12)^2 = 7.56$ and $F_{\text{critical}} = 5.41$ ($df_1 = 4 - 1 = 3$, $df_2 = 6 - 1 = 5$, Appendix D3), then since $F_{\text{calculated}} > F_{\text{critical}}$ the concentration of DDT to which birds are exposed to are significantly different in the two habitats.

b) Use Appendix D2 for t-value, the t-value at 90% (two-sided) is the same as that at 90% (one-sided). For Area 1, $t_{3,\ 0.95} = 2.353$, $\bar{x} = 1.2$, and the SEM $= (0.33/\sqrt{4} = 0.165)$. $\text{CI} = \bar{x} \pm t_{n-1,1-\frac{\alpha}{2}}\left(\frac{s}{\sqrt{n}}\right) = 1.2 \pm 2.353 \times 0.165 = 1.2 \pm 0.388$ ppb.

For Area 2, we have $t_{5,\ 0.95} = 2.015$, $\bar{x} = 1.8$, and SEM $= (0.12/\sqrt{6} = 0.0490)$. $\text{CI} = \bar{x} \pm t_{n-1,1-\frac{\alpha}{2}}\left(\frac{s}{\sqrt{n}}\right) = 1.8 \pm 2.015 \times 0.0490 = 1.8 \pm 0.099$ ppb.

c) First, we need to convert x value into standardized z value using $z = \dfrac{x - \mu}{\sigma}$. For $x = 2.2$, $z = 3.333$. For $x = 1.3$, $z = -4.167$. Use Appendix D1 for the P-value. $P(x > 2.2) = P(z > 3.333) = 1 - P(z < 3.333) = 1 - 0.9996 = 0.0004\,(0.04\%)$.

$$P(x<2.2) = P(z<3.333) = 0.9996\,(99.96\%)$$
$$P(1.3<x<2.2) = P(z<3.333) - P(z<-4.167) = 0.9996 - 0.0000 = 0.9996\,(99.96\%).$$

Note that Appendix D1 does not have the probability for a negative value of z. However, D1 can still be used because of the symmetric nature of the normal distribution curve. For example, $P(z < -3.333) = 1 - P(z > 3.333)$ because the probability for $z < -3.333$ is the same for probability for $z > 3.333$. Also, D1 does not give the probability value if z values are greater than 3.99, because $P(z<3.99) = 1.0000$, and $P(z < -3.99) = 0$. Thus, in the above calculation, $P(z < -4.167) = 0$.

15) **PCB concentrations in fish tissues in two rivers have been determined. In River A the concentrations (mg/kg) were as follows: 2.34, 2.66, 1.99, 1.91 (Mean $= 2.225, s = 0.3449$); in River B (mg/kg): 1.55, 1.82, 1.34, 1,88 (Mean $= 1.648, s = 0.250$). Determine the statistical significance of the concentration difference between the two rivers (a 90% confidence level assumed).**

$$F = s_1^{\,2} / s_2^{\,2} = (0.3449)^2 / (0.2502)^2 = 1.900$$

From Appendix D3, $F_{critical} = 9.28$ ($df_1 = 3$, $df_2 = 3$). $F < F_{critical.}$, Thus, PCB concentrations are not statistically significantly different.

16) **The USEPA Regional Screening Level (RSL) for soil arsenic under unrestricted use (e.g., residential) assumptions is 0.39 mg/kg. If one wishes to test whether the mean amount of arsenic in a soil exceeds the RSL, 0.39 mg/kg, (a) Formulate the appropriate null and alternative hypotheses. (b) Describe a Type I error and a Type II error.** The null hypothesis follows a "no difference" convention; thus, H_0: RSL for soil arsenic = 0.39 mg/kg. Since the hypothesis is directional, the alternative hypothesis test is one-tailed; thus, H_a: soil arsenic level > 0.39 mg/kg.

a) Type I error = false positive = EPA decides "soil is contaminated" but "in fact, it is not." A Type I error in this case will falsely report that the area exceeds 0.39 mg/kg of arsenic, while in fact it does not. This false-positive result favors the EPA rather than the responsible party.

b) Type II error = false negative = EPA decides "soil is not contaminated" but "in fact, it is." A Type II error will falsely report that the area has less than 0.39 mg/kg of arsenic, while in fact it exceeds the maximum allowable. This false-negative result is favorable to the responsible party rather than the EPA.

17) **The following calibration data were obtained for a total organic carbon (TOC) analyzer measuring TOC in water:**

Concentration, µg/L	Number of replicates	Mean signal (mv)	Standard deviation
0.0	20	0.03	0.008
6.0	10	0.45	0.0084
10.0	7	0.71	0.0072
19.0	5	1.28	0.015

a) **Do the regression analysis ($y =$ meansignal, $x =$ concentration) using Excel. Include at least the summary output, ANOVA, line plot, residual plot, and normal probability plot.**

b) **What is the best straight line by least squares fit? Indicate the slope and intercept of this linear fit. Indicate whether this linear regression is significant and whether the intercept and slope are significant. Use a significant level of 0.05.**

c) **Use Excel function SLOPE () and INTERCEPT() to confirm the answer from (b).**

d) **What is the calibration sensitivity in mv/(μg/L)?**

   a) The Excel output is attached, including the summary output, ANOVA, line plot, residual plot, and normal probability plot.

   b) The best fitted linear equation is: $y = 0.0656x + 0.0438$, where 0.0656 mv/(μg/L) is the slope and 0.0438 mv is the intercept. The significance of the linear relationship can be seen either from the F-value compared to $F_{critical}$, or the p-value of the slope. Since F (2946.093) > $F_{critical}$ (0.000339), the linear relationship is significant. Since the p-value for the slope = 0.000339 < 0.01, it is significant at a 99% confidence level.

   c) Using SLOPE () and (INTERCEPT () functions returns the same slope and intercept values as shown in the regression output (see also (b)).

   d) Analytical sensitivity increases as the slope of the calibration curve increases. The sensitivity equals to the slope, i.e., 0.0658 mv/(μg/L).

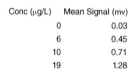

Conc (μg/L)	Mean Signal (mv)
0	0.03
6	0.45
10	0.71
19	1.28

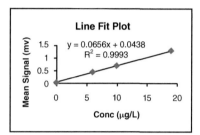

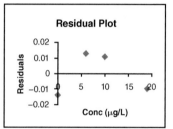

SUMMARY OUTPUT

Regression Statistics	
Multiple R	0.99966074
R Square	0.999321595
Adjusted R Sc	0.998982393
Standard Erro	0.016684569
Observations	4

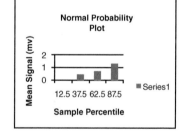

ANOVA

	df	SS	MS	F	Significance F
Regression	1	0.8201183	0.820118	2946.093	0.000339
Residual	2	0.0005567	0.000278		
Total	3	0.820675			

	Coefficients	Standard Error	t Stat	P-value	Lower 95%	Upper 95%	Lower 95.0%	Upper 95.0%
Intercept	0.043761468	0.0134658	3.249832	0.083058	−0.01418	0.1017	−0.01418	0.1017
Conc (μg/L)	0.065570118	0.001208	54.27793	0.000339	0.060372	0.070768	0.060372	0.070768

RESIDUAL OUTPUT

Observation	Predicted Mean Signal (mv)	Residuals	Standard Residuals
1	0.043761468	−0.013761	−1.01017
2	0.437182176	0.0128178	0.940903
3	0.699462647	0.0105374	0.773503
4	1.289593709	−0.009594	−0.70423

PROBABILITY OUTPUT

Percentile	Mean Signal (mv)
12.5	0.03
37.5	0.45
62.5	0.71
87.5	1.28

18) **A survey on the background concentration of lead in soil was conducted in an Environmental Impact Assessment of a former battery plant. Preliminary results of six soil samples in the protected area give results in mg/kg: 28, 32, 21, 29, 25, 22. (a) Is there any value in this data set that can be discarded as an outlier at 90% confidence level? (b) Calculate CV, RSD, and 95% confidence interval (CI). (c) If sufficient numbers of soil samples were taken and the population was known to have a normal distribution with a mean of 25.5 mg/kg and standard deviation of 1.2 mg/kg, what would be the probability of a soil sample having Pb concentration of greater than 29.8 mg/kg; less than 21 mg/kg; and in the range of 23.1 to 27.4 mg/kg?**

a) Using Grubbs' test, $T_{\text{critical}} = 1.672$. Possible outliers are the smallest and largest numbers (i.e., 21 and 32). For $x_1 = 21$, $T = (26.16667-21)/4.262237 = 1.212$. For $x_n = 32$, $T = (32 - 26.16667)/4.262237 = 1.369$. Since both 1.212 and 1.369 are smaller than $T_{\text{critical}} = 1.672$, there are no outliers in this data set. Below is the output from Excel's descriptive statistics.

Pb (ppm)	Pb (ppm)	
28		
32	Mean	26.16667
21	Standard error	1.740051
29	Median	26.5
25	Mode	#N/A
22	Standard deviation	4.262237
	Sample variance	18.16667
	Kurtosis	−1.47876
	Skewness	0.060269
	Range	11
	Minimum	21
	Maximum	32
	Sum	157
	Count	6
	Largest	32
	Smallest	21
	Confidence level (95%)	4.472936

b) $CV = s/\bar{x} = 4.262237/26.16667 = 0.1629$

$RSD = s/\bar{x} \times 100 = 16.29\%$

95% Confidence Interval (CI) $= \bar{x} \pm t_{n-1,1-\frac{\alpha}{2}}\left(\dfrac{s}{\sqrt{n}}\right) = 26.17 \pm 2.571 \times 4.26/\sqrt{6} = 26.17 \pm 4.47$

(*t*-value = 2.571 for two-sided 95% confidence). Note that the confidence interval of 4.47 can also be read directly from Excel's descriptive statistics.

c) First, we need to convert x value into standardized z value using $z = \dfrac{x-\mu}{\sigma}$ before we can use the table in Appendix D1 for the p-value (probability). For $x = 29.8$, 21, 23.1, and 27.4, the corresponding values of $z = 0.8524$, −1.2122, −0.7195, and 0.2894.

$P(x > 29.8) = P(z > 3.58) = 1 - P(z < 3.58) = 1 - 0.9998 = 0.0002\,(0.02\%)$.

$P(x < 21) = P(z < -3.75) = 1 - P(z > 3.75) = 1 - 0.9999 = 0.0001\,(0.01\%)$.

$P(23.1 < x < 27.4) = P(-2.00 < z < 1.58) = P(z < 1.58) - P(z < -2.00)$

$= P(z < 1.58) - (1 - P(z > 2.00)) = 0.9429 - (1 - 0.9772) = 0.9201\,(92.01\%)$.

19) **A new method is being developed for the analysis of a pesticide in soils. A spiked sample with a known concentration of 17.00 mg/kg was measured five times using the new method (data: 15.3, 17.1, 16.7, 15.5, 17.3 mg/kg) and the established EPA method (data: 15.4, 15.9, 16.7, 16.1, 16.2 mg/kg). The Excel outputs of descriptive statistics and one-way ANOVA are given below ($a = 0.05$). On the basis of these output, (a) are these two methods significantly different at a 95% confidence level? (b) Which method is more accurate, why? (c) Which method is more precise, why?**

1) Excel output of descriptive statistics

EPA method		New method	
Mean	16.06	Mean	16.38
Standard error	0.211187121	Standard error	0.412795349
Median	16.1	Median	16.7
Mode	#N/A	Mode	#N/A
Standard deviation	0.472228758	Standard deviation	0.923038461
Sample variance	0.223	Sample variance	0.852
Kurtosis	1.012889863	Kurtosis	−2.910136878
Skewness	−0.105406085	Skewness	−0.400799173
Range	1.3	Range	2
Minimum	15.4	Minimum	15.3
Maximum	16.7	Maximum	17.3
Sum	80.3	Sum	81.9
Count	5	Count	5
Largest	16.7	Largest	17.3
Smallest	15.4	Smallest	15.3
Confidence level (95%)	0.586350662	Confidence level (95%)	1.146106

2) Excel output of one-way ANOVA

Summary				
Groups	*Count*	*Sum*	*Average*	*Variance*
EPA method	5	80.3	16.06	0.223
New method	5	81.9	16.38	0.852

ANOVA						
Source of variation	*SS*	*df*	*MS*	*F*	*p-Value*	*F crit*
Between groups	0.256	1	0.256	0.476279	0.509634	5.317655
Within groups	4.3	8	0.5375			
Total	4.556	9				

a) We can use either *F*-test or one-way ANOVA for this problem.

F-test: $F = \dfrac{s_1^2}{s_2^2} = \dfrac{0.9203^2}{0.4722^2} = 3.798 < F_{critical} (= 6.39, df_1 = df_2 = 5-1 = 4$, Appendix D3), it is not statistically significantly different. Or:

One-way ANOVA: $F (= 0.476) < F_{critical} (= 5.317)$, it is not statistically significantly different. The *p*-value of 0.5096 > 0.05 also indicates that the difference of these two methods is not statistically significant.

b) The new method is more accurate since the mean of the new method (16.38) is closer to the true value (17.00). The EPA method is more precise, because its RSD of 2.94% $(0.4722/16.06 \times 100)$ is smaller than that of the new method $(0.9230/16.38 \times 100 = 5.63\%)$.

20) **Two analysts were given a standard reference material (SRM) and were asked to determine its copper concentration (in mg/kg) using the EPA standard method. Each analyst was given sufficient time so they could produce as many accurate results as possible. Their results (mg/kg) were as follows:**

Analyst A 45.2, 47.3, 51.2, 50.4, 52.2, 48.7 (standard error $s = 2.62$)
Analyst B 49.4, 50.3, 51.6, 52.1, 50.9 (standard error $s = 1.06$)

a) **For Analyst A, is 52.2 mg/kg a possible outlier?**
b) **Calculate the RSD for each student; which analyst is more precise?**
c) **If the known concentration of copper in SRM is 49.5 mg/kg, which analyst is more accurate (without considering the deletion of outliers if any)?**

a) Using Dixon's test, we first arrange the data in an increasing order: 45.2, 47.3, 48.7, 50.4, 51.2, 52.2. If the largest number 52.2 is a suspected outlier,

$D_{10} = \dfrac{x_6 - x_5}{x_6 - x_1} = \dfrac{52.2 - 51.2}{52.2 - 45.2} = 0.1429 < 0.560$ (Table 3.8, $n = 6$, risk of false rejection

= 5%). Hence, it is not an outlier. If the smallest number 45.2 is a suspected outlier, $D_{10} = \dfrac{x_2 - x_1}{x_6 - x_1} = \dfrac{47.3 - 45.2}{52.2 - 45.2} = 0.300 < 0.560$ (Table 3.8, $n = 6$, risk of false rejection

= 5%). Hence, it is not an outlier.

b) RSD(Analyst A)$= s/\bar{x} \times 100 = \dfrac{2.62}{49.167} \times 100 = 5.33\%$

RSD (Analyst B) $= s/\bar{x} \times 100 = \dfrac{1.06}{50.86} \times 100 = 2.08\%$

Analyst B is more precise because of the smaller RSD.

c) The means of the measured concentrations were 49.167 for Analysis A and 50.86 for Analyst B. Analyst A is more accurate because the measured mean concentration by Analyst A is closer to the true value of 49.5 mg/kg. Alternatively, Analysts A and B have recovery of 99.33% (= 49.167/49.5 × 100) and 102.75% (50.86/49.5 × 100). The recovery for Analyst A is closer to 100%, and he/she is more accurate.

21) **Use the Excel functions to calculate the probability (*P*) of normal function *f(x)* or standard normal function *f(x)*: (a) $P(z > 1.52)$, (b) $P(-1.23 < z < 2.57)$, (c) $P(x > 1.52)$ for $\mu = 1.32$, $\sigma = 1.5$, (d) $P(-1.23 < x < 2.57)$ for $\mu = 1.32$, $\sigma = 1.5$.**

a) NORMSDIST (1.52) = 0.9357, $P(z > 1.52) = 1 - P(-\infty < z < 1.52) = 1-0.9357 = 0.0643$ (6.43%).
b) NORMSDIST (−1.23) = 0.1093, NORMSDIST (2.57) = 0.9949 (99.49%)
$P(-1.23 < z < 2.57) = P(-\infty < z < 2.57) - P(-\infty < z < -1.23) = 0.9949 - 0.1093$
$= 0.8856$ (88.56%)

c) Note that for (a) and (b), the probability was for a normal distribution based on the normalized z value. Here the probability is based on the x value. We can convert x to z and then use Excel function NORMSDIST(z) for standard normal distribution. Alternatively, we can directly use NORMDIST(x, μ, σ, 1) without the conversion.

NORMDIST(1.52, 1.32, 1.5, 1) = 0.5530

$P(x > 1.52) = 1 - P(-\infty < x < 1.52) = 1 - 0.5532 = 0.4470$ (44.70%)

d) Similar to (c), we use NORMDIST(x, μ, σ, 1) function.

NORMDIST(2.57, 1.32, 1.5, 1) = 0.7977

NORMDIST(−1.23, 1.32, 1.5, 1) = 0.0446
$P(-1.23 < x < 2.57) = P(-\infty < x < 2.57) - P(-\infty < x < -1.23) = 0.7977 - 0.0446$
$= 0.7531 (75.31\%)$

22) **Find the critical value of: (a) t with $\alpha = 0.01$ and $n = 15$; (b) F with $\alpha = 0.01$, $n_1 = 13$ (larger variance), $n_2 = 10$.**

a) $df = n - 1 = 15 - 1 = 14$, TINV(α, df) = TINV(0.01, 14) = 2.977. Alternatively, the same critical t-value can also be obtained from the tabulated value in Appendix D2.

b) $df_1 = n - 1 = 13 - 1 = 12$, $df_2 = 10 - 1 = 9$, FINV(0.01, 12, 9) = 5.11. Note that df_1 is reserved for the sample group with a larger variance; otherwise, df_1 and df_2 need to be switched. Also, the tabulated critical F-values in Appendix D3 are only available for $\alpha = 0.05$ (i.e., 95% confidence). If this problem is misread as $\alpha = 0.05$, then $F_{critical} = 3.07$ from Appendix D3. At any other α or confidence levels, Excel functions FINV(α, df_1, df_2) should be used.

23) **Liu et al. (2022) reported the concentrations (μg/L) of PFPeS in contaminated groundwater in Canadian airports as follows: ND, ND, ND, 33.9, 48.3, 7.1, 0.4, 1.0, 5.0, 9.9, 1.7, ND, 2.8, 2.1 ($n = 14$). The LOQ was 0.084 μg/L. (a) Demonstrate why deletion of ND data points or the substitution of ND with zero and LOQ could be biased. (b) Suggest a method or methods to report this censored data.**

a) The descriptive statistics outputs from Excel for the deletion and three methods of substitutions are shown in the table below. The deletion of four ND data points results in a mean of 11.22, which is considerably higher than the means from the three substitution methods. This implies that the deletion of nondetected censored data most likely will result in a biased higher estimate. On the contrary, substitution of ND with zero leads to another extreme that could be biased low value of estimate.

b) Suggested methods for analyzing censored nondetected data include the Kaplan–Meier (K-M) method, regression on order (ROS) statistics, and maximum likelihood estimation (MLE) available from R. MLE is available from Minitab. Graduate students who have the access to R and Minitab are encouraged to explore them. None of these methods, unfortunately, are available from Excel (see Example 3.6).

	Delete	Zero	1/2DL	DL
Mean	11.22	8.014286	8.026286	8.038286
Standard error	5.177876	3.903096	3.901204	3.899318
Median	3.9	1.9	1.9	1.9
Mode	#N/A	0	0.042	0.084
Standard deviation	16.37388	14.60405	14.59697	14.58991
Sample variance	268.104	213.2782	213.0715	212.8655
Kurtosis	2.345519	4.523004	4.528066	4.533089
Skewness	1.828157	2.283667	2.285043	2.286406

(*continued*)

Range	47.9	48.3	48.258	48.216
Minimum	0.4	0	0.042	0.084
Maximum	48.3	48.3	48.3	48.3
Sum	112.2	112.2	112.368	112.536
Count	10	14	14	14
Confidence level (95.0%)	11.71317	8.432127	8.428039	8.423964

Chapter 4

Questions

1) **Why are the data quality objectives (DQOs) important prior to the implementation of environmental sampling?**

   DQOs are the "qualitative and quantitative statements that define the appropriate type of data and specify the tolerable levels of potential decision errors that will be used as basis for establishing the quality and the quantity of data needed to support a decision." Hence, DQOs are important to achieve *the least expensive data with the most certainty*.

2) **Define "representativeness" and discuss how sample "representativeness" is dependent on both the project objective and sample matrices. Give examples to illustrate.**

   Representativeness is defined as the measure of the degree to which data accurately and precisely represent a characteristic of a population, a parameter variation at a sampling point, a process condition, or an environmental condition.

   Sample "representativeness" depends on both project objective and the types of sample matrices (air, water, soil, biological samples, etc.). In the lagoon example (Fig. 4.4), it illustrates why samples A, B, and C would not be representative if the project goal is to assess the contamination of the entire area impacted by the discharge. Another example would be for air sampling, a distribution of sampling sites representing a worst-case scenario would not fit that of a best-case scenario.

   The dependence of sample representativeness on sample matrices is apparent. For example, chemical concentrations will probably vary considerably along the depths in soil. Water samples will vary depending on the seasons (water balance and precipitation, etc.), tidal influence, and stratification (lakes and deep and stagnant streams). Biological sample may have unique challenges because of the different species, size, sex, mobility, and tissue variations.

3) **Make a list of important factors (criteria) that are important in developing a sampling design including where, when, how many, and how frequently samples are collected.**

   In developing a sampling plan to formulate the number and location of samples (how many, where, and when), four primary factors need to be considered (Gilbert, 1987, Fig. 4.3). Of these four factors (objectives, variability, cost factors, and nontechnical factors), a project objective is probably the determining factor in sampling design. For example, sampling efforts in water quality monitoring will be very different depending on whether the objective is for trend analysis or baseline (background) type investigation. The former needs a long-term but less frequent

Fundamentals of Environmental Sampling and Analysis, Second Edition. Chunlong Zhang.
© 2024 John Wiley & Sons, Inc. Published 2024 by John Wiley & Sons, Inc.
Companion Website: www.wiley.com/go/EnvironmentalSamplingandAnalysis2e

sampling scheme, while the latter requires more samples and perhaps a one-time sampling event. The project objective with regard to the required data quality also affects the number of samples to be collected. In other words, the sample number will increase significantly as the allowable margin of error is reduced. Environmental variability or the spatial/temporal patterns of contamination is another important consideration (detailed in Section 4.2). Sampling approaches need to address such variations in order to obtain "representative" samples.

Expenses associated with sampling and analysis should certainly be a consideration in all environmental projects. The cost-effectiveness of a sampling design should be evaluated in the design stage so that the chosen sampling design will achieve a specified level of data quality at a minimum cost or, an acceptable level of data quality at a prespecified cost. Last but not least important is that sampling design should consider other factors such as sampling convenience, site accessibility, availability of sampling equipment, and political aspects. These nontechnical factors have a significant impact on the final sampling design. On the top of the above considerations, regulation constraints in most cases always need to be consulted first. A good example is the sampling related to remediation projects under the Superfund Act. The US EPA has specific guidelines for representative sampling of soil, air, biological tissue, waste, surface water, and sediment in Superfund investigation.

4) **Describe various reasons that can cause heterogeneity, that is, the difficulties in obtaining representative samples from (a) surface water (river, stream, lake) and (b) atmosphere.**

   a) Surface waters (rivers, streams, lakes): Surface water can be very heterogeneous both spatially and temporally as a result of seasonal flow variations (water balance, precipitation) and stratification (due to temperature and salinity gradients).

   b) Atmosphere: Chemical concentrations in the atmosphere at the same location may have significant differences within minutes depending on the changes in local meteorological conditions (i.e., wind velocity, direction).

5) **Describe various reasons that can cause heterogeneity, that is, the difficulties in obtaining representative samples from (a) solids (e.g., soils) and (b) biological samples (e.g., fishes).**

   a) Solids (e.g., soils): Contaminants in soils may have only slight temporal variations in a short term, but spatial variations along the depths of soils may be significant. Furthermore, on the microscopic scale, some contaminants may be in the form of "nugget," or may be preferentially adsorbed in certain smaller soil particles.

   b) Biological samples (fish): Obtaining a small but representative subsample for biological samples such as fish are particularly challenging. This is because of the variations of chemical concentrations due to the difference in biological species, size, sex, mobility, and various tissues.

6) **Give an example of how spatial scale of sampling can be dictated by the program objective(s).**

   Program objectives dictate sampling locations. For example, if the objectives are to assess compliance with air quality standards or assess long-term trends, the monitoring stations should be located where concentrations are expected to be largest or spatial concentration distributions can be estimated most accurately. If the objective is to provide data during episodes, the stations should be located where concentrations are expected to be largest. If the objective is to monitor source compliance with regulations or provide data to support enforcement actions, the stations should be located where the sensitivity of concentration levels to source emission level change is greatest. If the objective is to provide data for research, the stations should be located where the spatial distribution of concentrations can be estimated most accurately.

7) **Explain what emissions, meteorological, and other factors contribute to the observed spatial and temporal concentration changes in the city of Delhi, India (refer to Figs 4.5 and 4.6).**

The weekday vs. weekend patterns (Fig. 4.5) might be related to the difference in emission, whereas the seasonal average difference (Fig. 4.6) could reflect both emissions and meteorological factors. For example, the considerably higher $PM_{2.5}$ during the winter relates to the emission from the institutional and residential uses of fossil fuels and biomass for heating purposes. The monsoon season experiences the lowest $PM_{2.5}$ because the washout of particulate matter from atmosphere.

8) **By carefully examining the data in Fig. 4.5(a), does $PM_{2.5}$ show weekday-weekend patterns? Explain the likely reasons.**

Note that 9/1/2019, 9/8/2019, 9/15/2019, 9/22/2019 are all weekends. $PM_{2.5}$ concentrations on both sites in Delhi, India, appear to be lower during the weekend than on the weekdays, likely due to the reduced construction and automobiles on weekends compared to the weekdays.

9) **Describe the differences between haphazard sampling and simple random sampling.**

Haphazard sampling: Typically, this means samples are from convenient places and times. Samples are taken anywhere, at any time, or for any reason. There is no statistical basis for haphazard sampling, so there is little to no ability to defend this data in court.

Judgmental sampling: The samples are taken from places and times that are known to provide some needed information. These are typically provided by technical experts, process employees, or other information. This is not a statistically based sampling approach. In limited cases (such as identification of spill sites), this approach is valid in providing useful information, before we spend more resources to take more samples.

Simple random sampling: This approach is based on statistical randomization. Every location, place, time, or the equivalent has an equal opportunity to be selected. Thus, there is no human bias in selecting specific sample locations (temporal and/or spatial). This is defensible in a court of law.

10) **Define: (a) composite sampling; (b) transect sampling; and (c) search sampling.**
   a) Composite sampling: This is an approach through which one obtain large numbers of samples and mix them up into smaller numbers of samples based on their representations. It is usually used when sampling is cheap but analysis is not.
   b) Transect sampling: This is another variation of systematic grid sampling. It involves establishing one or more transect lines across a surface. Samples are collected at regular intervals along the line.
   c) Search sampling: This sampling approach utilizes either a systematic grid or systematic random sampling approach to define the minimum grid size in order to locate hot spots such as contaminated sites.

11) **Describe the advantages and disadvantages of (a) composite sampling and (b) systematic sampling.**
   a) Composite sampling: By reducing the number of samples for chemical analysis, composite sampling provides cost savings particularly for expensive-to-analyze trace chemicals. Physical mixing of subsamples does not change the average concentration of the analytes present in the samples, but information on the variations of the analytes is lost after the mixing. Also, composite sampling cannot be used for VOCs sampling.
   b) Systematic sampling: Systematic sampling (either grid or random) provides more uniform distribution (temporal or spatial), and thus helping to delineate, for example, how much contamination and define concentration gradients. The systematic grid sampling is easier to implement than simple random sampling and systematic random sampling.

12) **Describe (a) the difference between systematic grid sampling and systematic random sampling and (b) the method to calculate the mean and standard deviation from systematic sampling.**

   a) Systematic grid sampling subdivides the area of concern by using square or triangular grids and samples from the nodes (the intersections of the grid line) or a fixed location (e.g., center) of each grid. Systematic random sampling subdivides the area into the same square or triangular grids. However, samples with each grid are collected using random sampling. Systematic grid sampling is easier and more convenient to use (e.g., all from the centers of the grids).

   b) The method to calculate the mean and standard deviation from systematic grid/random sampling is the same as that in simple random sampling, i.e., $\bar{x} = \frac{1}{n}\sum x_n$; $s = \sqrt{\frac{\sum (x_i - \bar{x})^2}{n-1}}$

13) **Why the standard deviation is typically smaller for stratified random sampling than simple random sampling, particularly for a heterogeneous population with a geographical (spatial) or temporal pattern?**

   In stratified random sampling, the sample size can be adjusted depending upon the variations or the cost of sampling in various strata. Strata expected to be more variable are sampled more extensively. This provides greater precision than simple random sampling. This also means that a stratified random sampling results in a smaller standard deviation.

14) **In a site assessment for the identification of a contaminated source, which one of the following is likely the best and the least favorable: (a) judgmental sampling; (b) simple random sampling; and (c) systematic random sampling? Briefly explain.**

   Judgmental sampling is probably the most favorable for identification of contaminated sources. With professional judgment using prior knowledge and experience in the population to be sampled, one may get things done quickly and less expensively. For example, one may just take samples where spills can be visualized to confirm contamination from spills. On the other hand, simple random sampling is probably the least favorable. It applies to sites with little background information to use from, which typically leads to more samples need to be collected and thus a more expensive sampling campaign.

15) **To confirm whether a site has been cleaned up or not, which one of the following is likely the best and the least favorable: (a) judgmental sampling; (b) simple random sampling; and (c) systematic random sampling? Briefly explain.**

   Systematic random sampling is likely the most favorable for this purpose. With this approach, an area is equally divided into grids and samples with each grid are collected according to a specified pattern. For sampling in space, it is easier to implement and more convenient than simple random sampling. For temporal sampling, it is easier to administrate due to a fixed sampling schedule.

   Simple random sampling will probably be the next favorable option. However, this usually requires an increasing number of samples compared to systematic random samples. If the number of samples is insufficient relative to a large area, it is likely that these sampling locations may not represent the entire area to confirm whether it is cleaned up or not.

   The judgmental sampling may only work if we are able to visually observe contamination. It will become immediately apparent if a spot of contamination is positively identified. However, to confirm the cleanness of the entire site for a legal purpose, such as in the case of remediation projects, this approach is not good enough. This nonstatistical approach cannot sustain legal challenge in a court of law.

16) **A petroleum refinery plant discharges toxic chemicals into a river at unknown periodic intervals. (a) If water samples below the discharge pipe will be collected to estimate the weekly average concentration of the effluent at that point, is simple random, stratified random, or systematic sampling the best? Briefly discuss the advantages and disadvantages of these three designs for this situation. (b) If the data objective is to estimate the maximum concentration for each week of the year, is your choice of a sampling plan different than in (a) where the objective was to estimate the weekly mean? Briefly state why?**

   a) Systematic sampling will be more likely to collect samples to the same extent among all weeks of the sampling period. Hence, the weekly average concentration is more likely to be achievable. However, if the discharges are also present in systematic patterns (e.g., maximum discharges occur at mid-nights), this approach (e.g., sampling every day at 10:00 am) may lead to misrepresentative samples of the actual discharge. Simple random sampling will typically require more samples to be collected. The stratified random sampling approach might be better than simple random sampling if we know the patterns (strata) of the weekly discharge, for example, high discharge during the weekdays and the low discharge during the weekends.

   b) If a sufficient number of samples are collected (i.e., the frequency of sampling is adequate), then systematic random sampling is preferred to obtain data regarding maximum discharge. This requires knowledge of discharge patterns. For example, if more samples are collected during the low discharge period, then it will underestimate the discharge data.

17) **Given the following particular situation, identify which sampling design approach should be used and discuss why this is the most appropriate.**

   a) **If you are performing a screening phase of an investigation of a relatively small-scale problem, and you have budget and/or a limited schedule. Your goal is to assess whether further investigation is warranted that should include a detailed follow-up sampling.**

   b) **If you are estimating a population mean and you have known the existence of spatial or temporal patterns of the contaminant. Your goal is to increase the precision of the estimate with the same number of samples, or achieve the same precision with fewer samples and lower cost.**

   c) **If you are estimating a population mean and you have an adequate budget. In the meantime, you are also interested in knowing the information of a spatial or temporal pattern.**

   d) **If you are estimating a population mean and you have budget constraints. The analytical costs are much higher compared with sampling costs. Your goal is to produce an equally precise or a more precise estimate of the mean with fewer analyses and lower cost.**

   e) **If you are developing an understanding of where contamination is present and you have an adequate budget for the number of samples needed, your goal is to acquire coverage of the area of concern with a given level of confidence that you would have detected a hot spot of a given size.**

   f) **If you are developing an understanding of where contamination is present and you have an adequate budget for the number of samples needed, your goal is to acquire coverage of the time periods of interest.**

a) If this is a screening phase of an investigation, plus you have budget and/or a limited schedule, your goal is to assess whether further investigation is warranted that should include a detailed follow-up sampling. *Judgmental sampling* is preferred because it is based on professional judgment using prior knowledge and experience. One wants things done quickly with limited resources.

b) If you are estimating a population mean and you have known the existence of spatial and temporal patterns of the contaminant. Your goal is to increase the precision of the estimate with the same number of samples, or achieve the same precision with fewer samples and lower cost. In such a case, a *stratified random sampling* is appropriate because strata can be used on the basis of spatial and temporal patterns of the contaminant.

c) If you are estimating a population mean and you have an adequate budget. In the meantime you are also interested in knowing the information of the spatial or temporal pattern. In this case, *systematic grid sampling* will be able to delineate the spatial or temporal pattern. This requires more samples but is warranted by the adequate budget.

d) If you are estimating a population mean and you have a budget constraint. The analytical costs are much higher as compared to sampling, goal is to produce equally or more precise estimate of the mean with fewer analysis and lower cost. In this case, *composite sampling* is the best, because we can now mix subsamples, thereby reducing the number of samples and the analytical costs. Physical mixing of samples will lose the variation information, but the mean should not be changed.

e) If you are developing the understanding of whether contamination is present, have an adequate budget, your goal is to acquire coverage of the area of concern with a given level of confidence that you would have detected hot spot of a given size. In this case, a *search sampling* is preferred, which will enable us to acquire coverage of the area of concern with a given level of confidence and in the meantime detect hot spot of a given size.

f) If you are developing an understanding of where contamination is present and you have an adequate budget for a number of samples needed, your goal is to acquire the coverage of the time period of interest. Although a large number of samples will be needed, a *simple random sampling* will justify its use because the budget is now not a big concern.

18) **You are tasked with reviewing and providing input for the data quality objectives for a large, high-budget soil remediation effort following a VOC contamination incident. One of the DQOs is to determine the minimum and maximum concentration of VOCs present, and incremental sampling was chosen to meet this objective. What would your input be?**

Justifications may vary. Incremental sampling (IS) is a special type of composite sampling that relies on mixing of subsamples to analyze the composite. VOCs would be lost during mixing, so this would not be a good choice. Composite sampling is generally a good solution for resource-limited sampling plans, as fewer samples are required to be analyzed. Since the budget is high for this project, a more thorough sampling method can be chosen. Also, IS would only be effective if the DOQ was to determine average concentrations, mixing leads to the loss of information on any variation data in the samples.

Problems

1) **In a study on the numbers of fish killed at the cooling water intake screens, El-Shamy (1979) found that sampling more intensively during the months of peak fish abundance and variance (e.g., April, May, August) resulted in a more precise estimate than when sampling was evenly distributed throughout the year. The following table shows the results for one of the fish species Alewife (*Alosa preudoharengus*) impinged at the Nine Mile Point Nuclear Stations Unit 1 on the southeastern shore of Lake Ontario in the town of Scriba, NY. The minimum number of samples allocated is four, and the methods of April, May, and August were assigned to 30, 31, and 31 sampling days, respectively, in the optimum allocation based on monthly variance strata.**

Month	Pooled variance	Sample size Current program	Optimum allocation	Optimum allocation with 25% reduction in sampling
Jan	54,434	13	4	4
Feb	10,870	12	4	4
March	93,147	13	4	4
April	92,446,178	16	30	28
May	2,357,299,702	18	31	31
June	5,123,587	13	14	7
July	4,498,057	13	14	6
Aug	51,093,570	14	31	22
Sept	3,575,257	13	12	5
Oct	48,083	13	4	4
Nov	949,118	13	6	4
Dec	2,572,739	13	10	4
Annual sample size		164	164	123
Mean±2SE		4104±1317	4104±150	4104±279
Precision		32%	4%	7%

a) **What type of sampling design can be inferred for the current program?**
b) **What type of sampling design can be inferred for the two optimum allocation methods of sampling?**
c) **Explain the pros and cons of these three sampling methods.**

   a) The current program generally fits *systematic random sampling* design in time on monthly basis. Here, one year is subdivided into 12 months as the grids. Within each month, the sampling days are approximately evenly distributed but the sampling dates are randomly selected.

   b) These two methods fit *stratified random sampling* design with proportional allocation. The number of sampling days within each stratum (month) is proportional to the fish abundance in each month.

c) As can be seen from the summary table, precision is the best for optimal allocation with 164 sampling days. When sampling days were reduced to 123, the precision got lower, but it was still better than the current program with evenly distributed sampling days within each month. Thus, better sampling precision can be obtained with less sampling efforts.

2) **To keep the fundamental error below 15%, what is the required minimum soil mass with soil particle size of 2 mm in diameter?**

By rearranging Eq. 4.7, we can estimate the minimum soil mass:

$$m = \frac{20\ d^3}{FE^2} = \frac{20 \times 0.2^3}{0.15^2} = 7.1\ \text{g}$$

3) **Estimate the fundamental error when 25 g of soil sample is used for laboratory analysis. Assume soil particle size of 0.2 cm. If soil mass is doubled, what will be the fundamental error?**

For soil mass $m = 25$ g, $FE = \sqrt{20 \times \dfrac{0.2^3}{25}} = 8.0\%$.

If m is doubled to m $= 50$ g, $FE = \sqrt{20 \times \dfrac{0.2^3}{50}} = 5.7\%$.

4) **A small land-based diesel oil spill has contaminated 2.5 acres of land in a rural area. After remediation, 60 incremental soil samples are collected using incremental sampling to confirm the clean-up. If square grids are used to locate increments, what increment spacing should be used?**

$$2.5\ \text{acres} = 2.5\ \text{acres} \times 43560\ \text{ft}^2 / \text{acre} = 108{,}900\ \text{ft}^2$$

$$L = \sqrt{108{,}900\ \text{ft}^2 / 60} = 42.6\ \text{ft}.$$

5) **A former small pesticide manufacturing facility was surveyed for pesticide residues in surrounding soils. Historical data have shown that the pesticide is very stable in soil, concentration is in the range of 40–200 ppb with a standard deviation of 5 ppb.**

a) **If an error level of ± 2 ppb is acceptable, how many samples are needed to be 95% confident that the requirement is met?**

b) **If an area of 1km^2 is to be surveyed (see the figure below), design the locations using the method of simple random sampling. Use Excel to generate random numbers and use the coordinate as shown below (i.e., $x = 0, y = 0$ for the manufacturing facility, $x = -500 \sim 500; y = -500 \sim 500$). Attach the random number from your Excel output and plot an x–y scatter plot showing the locations of all samples calculated from (a).**

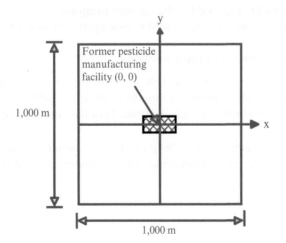

a) Error level $e = 2$ ppb, standard deviation $s = 5$ ppb. Since the t-value depends on the sample number n, this requires a trial-and-error method to find the minimal sample number n. Let us assume $n = 10$ (or any other number), then $df = 10 - 1 = 9$. This gives t value $= 2.262$ for a 95% confidence (Appendix D2).

$$n = \frac{t^2 s^2}{e^2} = \frac{2.262^2 \times 5^2}{2^2} = 31.97 \cong 32$$

If $n = 32$, then $df = 31$, $t = 2.042$ (Appendix D2), we now have the new value of n:

$$n = \frac{t^2 s^2}{e^2} = \frac{2.042^2 \times 5^2}{2^2} = 26.06 \cong 26$$

If $n = 26$, then $df = 25$, $t = 2.060$ (Appendix D2), we now have another new value of n:

$$n = \frac{t^2 s^2}{e^2} = \frac{2.060^2 \times 5^2}{2^2} = 26.52 \cong 27$$

On the conservative side, we round it up to $n = 27$ samples as the final answer.

b) The 27 pairs of random numbers (y vs x) are obtained using Excel's Random Number Generation feature under Data → Data Analysis → Random Number Generation. An example of the x–y scatter plot containing 27 sampling sites is shown below. Since the numbers are randomly generated, students' answers will vary.

x	y	
179.37	93.83	**Excel's Random Number Generation:**
84.31	164.88	Number of variables: 2
−493.13	10.39	Numbers of random numbers: 27
−430.97	212.49	Distribution: Uniform
243.40	8.59	Between: −500 and +500
414.24	−231.13	
464.51	176.53	
−89.25	−204.25	
281.55	−309.53	
499.15	−435.94	
−395.14	149.89	
−110.98	−80.28	
471.65	219.35	
259.36	308.95	
94.20	107.90	
−55.71	345.06	
285.55	−82.02	
−488.07	197.07	
418.00	394.59	
−357.91	−417.91	
−428.83	105.82	
−291.28	−357.60	
25.86	−413.05	
473.97	488.49	
−240.35	477.90	
−200.31	−310.60	
100.91	186.57	

6) **A pigment manufacturing process has been generating waste for a number of years. The wastes were discharged into a 40-acre lagoon. The pigment is generated in large batches that involve a 24-h cycle: 16 h of high percentage of large-sized black particulate matter and 8 h of smaller white pigment particles that are much smaller, settle more slowly and travel further toward the effluent pipe. The lagoon has distinct color strata, with the black and the gray sludge each covering a quadrant measuring 1320 ft by 330 ft, and the white sludge covering the remaining area of the lagoon, which measures 1320 ft × 660 ft. The leachable contaminant (barium) was assumed to be associated with the black sludge, which was concentrated in the first quadrant. The sludge had settled to a uniform thickness throughout the lagoon and was covered with 2 ft of water. Design a sampling plan (locations) using stratified random sampling. Assume that we need to collect a total of 10 samples from the black sludge area (heavily contaminated), 10 samples from the gray area (mixed, less contaminated), and 20 samples from the large strata (least contaminated).**

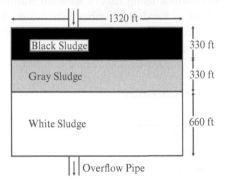

The random numbers shown below are generated using Excel (see below). The scatter plot is also attached. Note that all numbers are in reference to the original point (0, 0) located in the bottom left corner of the white sludge area. Students' answers will vary because of the randomness of the numbers.

	Black sludge		Gray sludge		White sludge	
Number of variables:	2		2		2	
Numbers of random numbers:	10		10		20	
Distribution:	Uniform		Uniform		Uniform	
Between:	x	y	x	y	x	y
	0 – 1320	990 – 1320	0 – 1320	660 – 990	0 – 1320	0 – 660
Random numbers:	x	y	x	y	x	y
	597	1049	687	781	260	409
	863	1148	571	715	363	225
	1084	1222	1222	883	850	22
	50	1045	316	777	1022	361
	832	1097	1278	804	24	161
	580	1272	173	692	1106	591
	1073	1271	35	682	1145	562

127	1260	500	917	505	227
146	1109	160	948	401	9
989	1304	332	710	1063	253
				22	116
				274	326
				978	498
				754	593
				52	331
				666	434
				1007	380
				976	226
				1222	615
				545	414

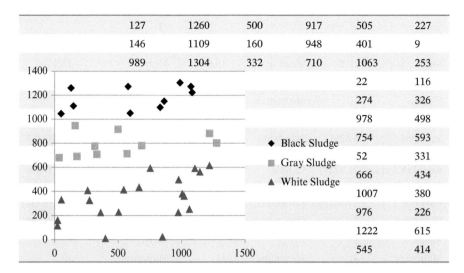

7) **Thermal stratification is common in lakes located in climates with distinct warm and cold seasons. It divides lakes into three zones (top: epilimnion; middle: thermocline; bottom: hypolimnion). Because of the stratification, the vertical mixing of water is prohibited. A stratified random sampling is designed to collect water samples for nitrogen concentrations. The following data were obtained:**

Stratum	Volume ($\times 10^6$ gal)	Number of samples taken	Nitrogen concentration (mg/L)
Epilimnion (0–8 ft)	5	8	8, 6, 11, 16, 9, 17, 4, 13
Thermocline (8–10 ft)	1	2	11, 16
Hypolimnion (10–25 ft)	14	10	25, 17, 18, 11, 12, 10, 22, 16, 10, 11

a) **One of the objectives was to estimate the mean, standard deviation, and confidence interval of the entire lake based on this stratified random sampling plan. Use 80% confidence level.**

b) **If the above nitrogen concentrations were obtained from simple random sampling (i.e., total number of samples = 8 + 2 + 10 = 20), calculate the mean, standard deviation, and confidence interval at an 80% confidence level.**

c) **Compare the results from (a) and (b) and comment on the difference.**

a) First, we use Excel to calculate the mean and standard deviation for each stratum (see below).

Epilimnion	Thermocline	Hypolimnion
8	11	25
6	16	17
11		18
16		11
9		12
17		10
4		22
13		16
		10

					11	
Epilimnion			*Thermocline*		*Hypolimnion*	
Mean	10.5		Mean	13.5	Mean	15.2
Standard error	1.636634		Standard error	2.5	Standard error	1.6786238
Median	10		Median	13.5	Median	14
Mode	#N/A		Mode	#N/A	Mode	11
Standard deviation	4.6291		Standard deviation	3.5355339	Standard deviation	5.3082745
Sample Variance	21.42857		Sample Variance	12.5	Sample Variance	28.177778
Kurtosis	−1.18272		Kurtosis	#DIV/0!	Kurtosis	−0.5943521
Skewness	0.138256		Skewness	#DIV/0!	Skewness	0.7630489
Range	13		Range	5	Range	15
Minimum	4		Minimum	11	Minimum	10
Maximum	17		Maximum	16	Maximum	25
Sum	84		Sum	27	Sum	152
Count	8		Count	2	Count	10
Largest	17		Larges	16	Larges	25
Smallest	4		Smallest	11	Smallest	10
Confidence level (95.0%)	3.870025		Confidence level (95.0%)	31.765512	Confidence level (95.0%)	3.7973108

Epilimnion: $k = 1$, $w_1 = 5 \times 10^6/(5 \times 10^6 + 1 \times 10^6 + 14 \times 10^6) = 0.25$, $\bar{x}_1 = 10.5$, $s_1 = 4.6291$

Thermocline: $k = 2$, $w_2 = 1 \times 10^6/(5 \times 10^6 + 1 \times 10^6 + 14 \times 10^6) = 0.05$, $\bar{x}_2 = 13.5$, $s_2 = 3.5355$

Hypolimnion: $k = 3$, $w_3 = 14 \times 10^6/(5 \times 10^6 + 1 \times 10^6 + 14 \times 10^6) = 0.70$, $\bar{x}_3 = 15.2$, $s_3 = 5.3083$

Then we can calculate the mean and standard deviation for the "entire" lake based on this stratified random sampling:

$$\bar{x} = \sum_{k=1}^{3} w_k \times \bar{x}_k = (0.25 \times 10.5 + 0.05 \times 13.5 + 0.70 \times 15.2) = 13.94 \text{ mg/L}$$

$$s^2 = \sum_{3}^{k=1} \frac{w_k^2 \times s_k^2}{n_k} = \left(\frac{0.25^2 \times 4.6291^2}{8} + \frac{0.05^2 \times 3.5355^2}{2} + \frac{0.70^2 \times 5.3083^2}{10} \right) = 1.564$$

$$s = \sqrt{s^2} = \sqrt{1.564} = 1.251$$

$$CI = \bar{x} \pm t \times \frac{s}{\sqrt{n}} = 13.94 \pm 1.328 \times \frac{1.251}{\sqrt{20}} = 13.94 \pm 0.37 \text{ (Appendix D2: when } n = 20, \text{ or } df = 19$$

at an 80% confidence level, $t = 1.328$)

b) If the same entire data were obtained using simple random sampling, then:

$$\bar{x} = \frac{8 + 6 + 11 + \ldots 22 + 16}{20} = 13.15 \text{ mg/L}$$

$$s = \sqrt{\frac{(8 - 13.15)^2 + (6 - 13.15)^2 + \ldots (16 - 13.15)^2}{20 - 1}} = 5.204$$

$$CI = \bar{x} \pm t \times \frac{s}{\sqrt{n}} = 13.15 \pm 1.328 \times \frac{5.204}{\sqrt{20}} = 13.15 \pm 1.54 \text{ (Appendix D2: when } n = 20, \text{ or } df = 19 \text{ at}$$

an 80% confidence level, $t = 1.328$)

The values of $\bar{x}$, s and CI can also be obtained directly from the Excel output using descriptive statistics (see below).

Raw data	Simple random sampling	
8		
6	Mean	13.15
11	Standard error	1.163649
16	Median	11.5
9	Mode	11
17	Standard deviation	5.203996
4	Sample variance	27.08158
13	Kurtosis	0.198842
11	Skewness	0.502661
16	Range	21
25	Minimum	4
17	Maximum	25
18	Sum	263
11	Count	20
12	Largest	25
10	Smallest	4
22	Confidence level (80.0%)	1.54501

c) As summarized below, the means are close. A slightly higher mean from stratified random sampling is the result of higher weight placed on the hypolimnion stratum which has higher nitrogen concentrations. However, the difference in the standard deviations is much smaller in stratified random sampling than in simple random sampling.

	Mean ($\bar{x}$)	Standard deviation (s)	Confidence interval (CI) at 80% confidence level
Stratified random sampling	13.94 mg/L	1.251 mg/L	13.94 ± 0.37 mg/L
Simple random sampling	13.15 mg/L	5.204 mg/L	13.15 ± 1.54 mg/L

8) **A stratified random sampling plan was adopted for a contaminated site as a result of a recent oil spill in an open agricultural land. Three strata were chosen as shown in the diagram below: (1) the heavily contaminated surface soil (0–2 ft); (2) the unsaturated soil (2 ft to the groundwater level at 7 ft deep); and (3) the saturated zone (7–27 ft). The results of a probe chemical (benzene) are shown in the table below. If one of the objectives was to estimate the degree of contamination in the entire aquifer (0–27 ft):**

a) **Estimate the overall mean, standard deviation, and confidence interval (at an 80% confidence level) using proportional allocation method (i.e., based on the depth or volume ratio).**

b) **Estimate the overall mean, standard deviation, and confidence interval (at an 80% confidence level) using optimal allocation method (i.e., based on the variation of each stratum).**

Stratum	Depth (ft)	Number of samples taken	Benzeneconcentration (mg/kg)
Surface soil (0–2 ft)	2	15	$\bar{x}_1 = 25; s_1 = 4.0$
Unsaturated soil (2–7 ft)	5	6	$\bar{x}_2 = 17; s_2 = 2.0$
Saturated soil (7–27 ft)	20	5	$\bar{x}_3 = 10; s_3 = 1.0$

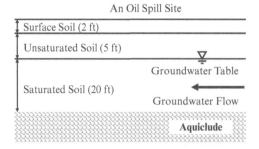

An Oil Spill Site

Surface Soil (2 ft)

Unsaturated Soil (5 ft)

Groundwater Table

Saturated Soil (20 ft)

Groundwater Flow

Aquiclude

a) Estimate the overall mean, standard deviation, and confidence interval (at an 80% CI) using **proportional allocation** method (i.e., based on the depth or volume ratio)

Surface soil: $k = 1$, $w_1 = 2/(2 + 5 + 20) = 0.074$, $\bar{x}_1 = 25$, $s_1 = 4.0$

Thermocline: $k = 2$, $w_2 = 5/(2 + 5 + 20) = 0.185$, $\bar{x}_2 = 17$, $s_2 = 2.0$

Hypolimnion: $k = 3$, $w_3 = 20/(2 + 5 + 20) = 0.741$, $\bar{x}_3 = 10$, $s_3 = 1.0$

Then we can calculate the mean and standard deviation based on stratified random sampling:

$$\bar{x} = \sum_{k=1}^{3} w_k \times \bar{x}_k = (0.074 \times 25 + 0.185 \times 17 + 0.741 \times 10) = 12.41 \text{ mg/kg}$$

$$s^2 = \sum_{3}^{k=1} \frac{w_k^2 \times s_k^2}{n_k} = \left(\frac{0.074^2 \times 4^2}{15} + \frac{0.185^2 \times 2^2}{6} + \frac{0.741^2 \times 1}{5} \right)^2 = 0.138$$

$$s = \sqrt{s^2} = \sqrt{0.138} = 0.372 \text{ mg/L}$$

$$CI = \bar{x} \pm t \times \frac{s}{\sqrt{n}} = 12.41 \pm 1.316 \times \frac{0.372}{\sqrt{26}} = 12.41 \pm 0.096 \text{ mg/L (Appendix D2: when } n = 26 \text{, or}$$

$df = 25$ at an 80% confidence level, $t = 1.316$)

b) Estimate the overall mean, standard deviation, and confidence interval (at an 80% confidence level) using **optimal allocation** method (i.e., based on variation of each stratum). Consider a special case when sampling costs are the same for all three strata, then the optimal allocation is the case termed Neyman allocation (Eq. 4.5).

$$n_k = n \frac{w_k \times s_k}{\sum_{k=1}^{r} w_k \times s_k}$$

$$n_1 = 26 \times \frac{0.074 \times 4}{0.074 \times 4 + 0.185 \times 2 + 0.741 \times 1} = 26 \times 0.210 = 5.46 \approx 5$$

$$n_2 = 26 \times \frac{0.185 \times 2}{0.074 \times 4 + 0.185 \times 2 + 0.741 \times 1} = 26 \times 0.263 = 6.84 \approx 7$$

$$n_3 = 26 \times \frac{0.741 \times 1}{0.074 \times 4 + 0.185 \times 2 + 0.741 \times 1} = 26 \times 0.527 = 13.70 \approx 14$$

$$s^2 = \sum_{3}^{k=1} \frac{w_k^2 \times s_k^2}{n_k} = \left(\frac{0.074^2 \times 4^2}{5} + \frac{0.185^2 \times 2^2}{7} + \frac{0.741^2 \times 1}{14} \right)^2 = 0.0763$$

$$s = \sqrt{s^2} = \sqrt{0.0763} = 0.276 \text{ mg/kg}$$

$$CI = \bar{x} \pm t \times \frac{s}{\sqrt{n}} = 12.41 \pm 1.316 \times \frac{0.276}{\sqrt{26}} = 12.41 \pm 0.071 \text{ mg/kg (Appendix D2: when } n = 26,$$

or $df = 25$ at an 80% confidence level, $t = 1.316$)

9) **A field-scale remediation demonstration project was studied to test the efficiency of soil vapor extraction to remove BTEX compounds from contaminated soils in the unsaturated zone due to a rupture of chemical container. It was estimated that the contaminated area should be within the boundary of 100 m (north-south) × 150 m (east-west). After the demonstration project is completed, a detailed sampling plan needs to be executed to precisely map the spatial pattern of BTEX concentrations. Budgetary constraints allow for 50 samples to be collected for BTEX analysis of soil cores. (a) If systematic grid sampling was employed, determine the grid spacing (L) using square grid. (b) Draw a schematic diagram showing how these 50 sampling points are located.**

   a) To determine grid spacing, use the formula: $L = \sqrt{A/n}$, where L is the grid spacing, A is the area to be sampled, and n is the total number of samples to be collected).

   $$L = \sqrt{\frac{A}{n}} = \sqrt{\frac{100 \times 150}{50}} = 17.3 \text{ m}$$

   b) The schematic shown below is systematic grid sampling. As shown, all 54 (for convenience, not the exact 50) sampling points are located in the center of each square grid.

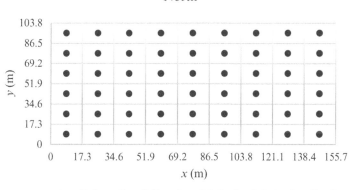

North

10) **A lagoon waste pit has the following historical data for the barium concentration based on a simple random sampling ($n = 4$): 86, 90, 98, 104 mg/kg (the lower two-thirds of lagoon). The regulatory threshold for barium is 100 mg/kg. The waste on this site was categorized to be hazardous, and therefore a more thorough sampling plan is needed. Determine the number of samples required so that the reported mean has a 90% confidence level.**

   First, calculate the mean and standard deviation using Excel or the following formulas:

   $$\bar{x} = \frac{86 + 90 + 98 + 104}{4} = 94.5 \text{ mg/kg}$$

   $$s = \sqrt{\frac{(86 - 94.5)^2 + (90 - 94.5)^2 + (98 - 94.5)^2 + (104 - 94.5)^2}{4 - 1}} = 8.06$$

Now we can apply Eq. 4.12 with $e = RT - \bar{x}$ to calculate the required sample number:

$$n = \frac{t^2 \times s^2}{\left(RT - \bar{x}\right)^2}$$

Here $n = 4$, $df = 4-1 = 3$, $t = 2.353$ at 90% confidence level (Appendix D2), regulatory threshold $(RT) = 100$ mg/kg, mean = 94.5. By substituting these values into the above equation, we can calculate the required number of samples:

$$n = \frac{2.353^2 \times 8.06^2}{\left(100 - 94.5\right)^2} = 11.9 \cong 12$$

Chapter 5

Questions

1) **Give a list of water quality parameters that must be measured immediately in the field while a sample is being taken.**

   These common water quality parameters include, but are not limited to, pH, salinity, Cl_2, ClO_2, CO_2, I_2, O_3, dissolved oxygen (by electrode), and temperature. They must be measured in the field immediately after the sample is taken.

2) **Which one of the following should be monitored in the field: (a) acidity, (b) DO, (c) pesticide, (d) hardness.**

   Only dissolved oxygen (DO) with the electrochemical probe method should be monitored in the field. The Winkler method requires the fixation step done on-site, but the subsequent titration can be done within eight hours of the sample collection.

3) **Describe all possible physical, chemical, and biological changes that can occur for the following compounds during the sample holding period: (a) oil, (b) S^{2-}, (c) CN^-, (d) phenols.**

   (a) Oil: adsorption to plastic container wall, biodegradation by bacteria. (b) S^{2-}: Volatilization via H_2S under acidic conditions. (c) CN^-: Volatilization via HCN under acidic conditions, and chemical reaction with Cl_2. (d) Phenols: Bacterial degradation.

4) **Describe all possible physical, chemical, and biological changes that can occur for the following compounds during the sample holding period: (a) chlorinated solvents, (b) PAHs, (c) metals.**

   (a) Chlorinated solvents: Volatilization because of their high volatility. (b) PAHs: Photochemical degradation. (c) Metals: Adsorption to glass wall, and precipitation in the forms of metal oxides and metal hydroxides under alkaline pH conditions.

5) **Explain why: (a) HNO_3 rather than other acids is used for metal preservation; (b) amber bottles are preferred for PAHs; (c) zero-headspace container for groundwater samples collected for tetrachloroethylene analysis.**

   (a) HNO_3 is commonly used for the preservation of samples for metal analysis, because salts of all metal nitrates are water soluble. Other acids such as sulfuric acid, hydrochloric acid, and phosphoric acid will likely precipitate certain metals from the solutions. (b) Amber bottles are preferred for PAH because the aromatic (double bond) nature of PAHs can readily absorb light and undergo photolysis. (c) Zero-headspace containers are needed for groundwater samples used for tetrachloethylene (PCE) analysis because PCE is a very volatile chlorinated compound. Any headspace should be eliminated by filling water sample to the full extent of the container.

Fundamentals of Environmental Sampling and Analysis, Second Edition. Chunlong Zhang.
© 2024 John Wiley & Sons, Inc. Published 2024 by John Wiley & Sons, Inc.
Companion Website: www.wiley.com/go/EnvironmentalSamplingandAnalysis2e

6) **What special precautions do you need to be aware of regarding sampling for VOCs analysis for each of the following: (a) sampling sequence relative to other analytes; (b) sample container; (c) composite sampling.**

(a) Water samples for VOCs analysis should always be collected first before we can proceed to collect samples for other parameters such as SVOCs, total metals, dissolved metals, etc. (b) For VOCs, there should not be any headspace in the sample container. Samples should be discarded if air bubbles are present in the container. (c) Composite sampling through mixing of subsamples for subsequent VOCs analysis is not permitted. Some of the VOCs may be lost due to volatilization during the mixing step.

7) **Explain potential contaminants that may leach from sample containers or sampling tools made of: (a) glass, (b) PVC, (c) plastics, (d) stainless steel.**

(a) Glass may leach certain amounts of boron and silica, and a significant sorption of metal ions may take place on the container wall. (b) PVC, in the case of threaded joints, may leach chloroform. It was reported that PVC-cemented joints may leach methyl ethyl ketone, toluene, acetone, methylene chloride, benzene, ethyl acetate, tetrahydrofuran, cyclohexanone, organic Sn compounds, and vinyl chloride. (c) Certain plastics may leach organic compounds (e.g., bisphenol A). (d) Stainless steels are not inert; they may leach certain metals including Cr, Fe, Ni, and Mo.

8) **Explain why glass containers are generally used for organic compounds, whereas PVC-type containers are used for inorganic compounds.**

Glass may leach certain amounts of boron and silica, and a significant sorption of metal ions may take place on the container wall. PVC-threaded joints may leach chloroform, while PVC-cemented joints may leach other organics such as methyl ethyl ketone, toluene, acetone, methylene chloride, benzene, ethyl acetate, tetrahydrofuran, cyclohexanone, organic Sn compounds, and vinyl chloride. It is, therefore, a common practice to use glass containers for organic compounds and use plastic containers for inorganic compounds whenever possible.

9) **Describe the difference between the Kemmerer sampler and the von Dorn sampler. Which one is particularly suitable for estuarine water sampling?**

A Kemmerer sampler can be lowered vertically into the water body at a desired depth and the sampler's body remains in a vertical position (valves open on both ends), then a messenger is sent down along the rope and hits the trigger to close the valves. A van Dorn sampler is a slight variation of Kemmerer in that it can be lowered into the water at a desired depth, where its body and both valves remain horizontal. It is more appropriate for estuary sampling (with salinity profile along the depths) or water bodies (such as lakes) with stratification.

10) **Which one of the following is used in the sampling of hazardous waste: (a) Ekman dredge, (b) Kemmerer, (c) Ponar, (d) Coliwasa.**

Coliwasa (composite liquid waste sampler) is used to collect liquid hazardous wastes, whereas Ekman and Ponar dredges are the sediment samplers and Kemmerer is a water sampler.

11) **The maximum holding time for metal analysis (excluding Cr^{6+} and dissolved Hg) after acidification to pH < 2 is most likely: (a) 6 h, (b) 6 days, (c) 6 weeks, (d) 6 months.**

A 6-month storage time is the maximum holding time (MHTs) for total metal analysis. The analysis of metals for certain species (such as Cr^{6+} and dissolved Hg) has a much shorter MHT.

12) **Which of the following is used to preserve samples for the analysis of total metals (excluding Cr and Hg): (a) Cool, $4^{o}C$, (b) NaOH to pH > 12, (c) H_2SO_4 to pH < 2, (d) HNO_3 to pH < 2?**

The addition of concentrated HNO_3 to keep water sample pH < 2.0 will preserve metals (excluding Cr and Hg). Cooling at $4°C$ is required for most other parameters, but it is not essential for metal samples.

13) **A "thief" is a two-concentric tube tool used to collect: (a) granular, particulate, or powdered waste, (b) liquid waste, (c) sticky and moist waste, (d) hard soils.**
A "thief" can be used to collect granular, particulate, and powdered waste. It is also called "grain sampler." A "trier" is used to collect sticky and loosened soils. It is a stainless-steel tube, cut in half lengthwise with a sharpened tip.

14) **Which one of the following is used to collect air samples for the analysis of low concentrations of organic contaminants: (a) TSP sampler, (b) PM_{10} sampler, (c) PS-1 sampler, (d) impinger?**
The PS-1 sampler, which draws air samples through polyurethane foam (PUF), or a combination of foam or adsorbing materials, is used to collect air samples for the analysis of low concentrations of organic contaminants. TSP and PM-10 samplers are used to collect total suspended solids (TSP) and particulate matter with a diameter of less than 10 μm, respectively. An impinger is used to collect compounds in condensable concentrations or those that can be readily retained by absorption or reactions with the liquid in the impinger.

15) **Which of the following sampling device/media is based on absorption and/or chemical reaction: (a) impinger, (b) XRD sorbent, (c) Tedlar bag, (d) canister?**
An impinger is used to collect compounds in condensable concentrations or those that can be readily retained by absorption or reactions with the liquid in the impinger. XRD sorbent retains analytes based on adsorption. Tedlar bags and canisters collect air samples in a container for subsequent withdrawal and chemical analysis.

16) **A bailer is commonly used to collect: (a) surface water sample, (b) groundwater well sample, (c) liquid hazardous waste sample, (d) none of the above.**
A bailer is a pipe with an open top and a check valve at the bottom (3 ft long with a 1.5-inch internal diameter and 1 L capacity). It is used to retrieve water samples from groundwater wells.

17) **Which of the following is the best for VOCs sampling from a groundwater well: (a) bailer, (b) air-lift pump, (c) peristaltic pump, (d) bladder pump? Explain.**
A bladder pump is recommended because it causes minimum alteration to a sample's integrity. A bladder pump uses compressed air to move the groundwater to a Teflon bladder enclosed by a stainless-steel or Teflon housing. It is suitable for VOCs. A peristaltic pump is suitable for sampling wells of a smaller diameter, but it is not suitable for VOCs because of possible VOC loss due to sample aeration. An airlift pump operates by releasing compressed air into the water; it is used only for well development rather than sampling.

18) **Suppose you were recently hired as an entry-level environmental field specialist in a new firm dealing with groundwater remediation where a SAP has not been developed. You are assigned as an assistant to a project manager for groundwater sampling, and are asked to do the office preparation for this sampling event. Make a list of items you may need in the field. Exclude containers and sampling tools from this list.**
Students can refer to the list of an "Ideal Tool Kit" (Bodger, 2003; textbook, page 136) for the commonly needed items. These include, but are not limited to, tools (wrench, screwdrivers, plier, hammer, duct tape, wire stripper, tape measure, flashlight, mirror, hand cleaner, gloves, paper towel, marker), decontamination supplies (e.g., brushes, plastic sheeting, bucket, solvent spray, aluminum foil, soap, and distilled water), waste disposal (e.g., trash bags and liquid waste containers), health and safety equipment (e.g., safety glasses, hard hat, steel-toe boots, first-aid kit, and ear plugs), and items for communication and documentation (e.g., cell phone, map, GPS, camera, chain-of-custody form). Also, have important telephone numbers and emergency contacts handy.

19) **Describe the special considerations in taking water samples from: (a) flowing waters (rivers and streams), (b) static waters (lakes and ponds), and (c) estuaries.**

   a) For fast-flowing waters (rivers and streams), sometimes it may be difficult to collect mid-channel samples. Health and safety concerns must dictate where to collect samples. When sampling near a point of pollution source, two samples from channel mid-depth are typically drawn. One upstream and one adjacent to, or slightly downstream of the point of discharge.

   b) For static waters, such as large lakes and larger impoundments, stratification due to temperature variations is often present. It is therefore important to collect several vertical aliquots.

   c) For estuaries, when water depth is less than 10 ft, samples are collected mid-depth unless the salinity profile indicates the presence of salinity stratification. In estuaries where water depth is >10 ft, water samples may be collected at 1 ft depth, mid-depth and 1 ft from the bottom. van Dorn sampler is often used rather than Kemmerer sampler.

20) **Describe why purging is important prior to groundwater sampling. What are the common approaches in groundwater purging.**

   Purging is important prior to groundwater sampling to remove stagnant water in the borehole and adjacent sandpack. After purging, the water that will be sampled should be stable and a representative sample can be obtained. The most common approach to groundwater purging is pumping the well until water quality parameters including pH, temperature, and conductivity are stabilized. "Stabilized" can be determined by *in situ* measurement of dissolved oxygen, turbidity, specific conductivity, oxidation-reduction potential and temperature.

21) **Explain why cross-contamination is especially important in groundwater sampling as compared to, for example, surface water sampling.**

   Cross-contamination is especially important in groundwater sampling because there are several potential contamination sources, including contamination during well construction, well purging, and use of sampling-related equipment. Unlike simple surface water sampling, one can't just directly collect well water samples. One needs to first purge the stagnant water until the water is stabilized. More equipment (i.e., sources of contaminants) is required for groundwater sampling than surface water sampling, including pumps, tubing, containers, monitoring probes, etc.

22) **Explain which one of the following has the least potential for cross-contamination: (a) PVC, (b) FRE, (c) Teflon, and (d) stainless steel.**

   Teflon has the least potential for cross-contamination because it is most inert and does not absorb metals or organics. Teflon also does not leak any detectable contaminants into the sample. Polyvinyl chloride (PVC), fiberglass-reinforced epoxy (FRE), and even stainless steel may leach some chemicals under certain conditions.

23) **Which surface sampling approach is appropriate in a tiled floor suspected of PCB contamination: (a) wipe sampling, (b) chip sampling, or (c) dust sampling?**

   Wipe sampling is appropriate for collecting potential PCBs on smooth and nonporous surfaces such as a tiled floor, wall, ventilation ducts and fans, etc. Chip sampling is usually performed with a hammer and chisel or with an electric hammer for chemical residues on porous surfaces such as cement, brick, or wood.

24) **Explain why passive samplers often improve the detection limit for organic compounds in water.**

   Passive sampling provides time-weighted concentration and because the polymer accumulates analytes of interest over time, the detection limit is much improved. This is in significant contrast to the traditional grab sampling that only represents a snapshot of the contaminant level at the time of sampling.

25) **Describe the difference between SPMDs and POCIS with respect to their applicable contaminants.**

SPMDs consist of lay-flat tubing made of low-density polyethylene (LDPE) filled with a high molecular weight lipid (triolein). Compounds with log $K_{ow} > 3.0$ are ideal for trapping in triolein, such as hydrophobic semivolatile organic compounds, PAHs, PCBs, and organotin compounds in air, water, sediments, and wastewater. Large macromolecules, ionic compounds and polar compounds do not dissolve into the membrane.

POCIS contains a mixture of three solid phase sorbents (Isolute ENV, polystyrene divinyl benzene, and Ambersorb 1500 carbon) dispersed on S-X3 biobeads. Unlike SPMD, POCIS concentrates waterborne hydrophilic (polar) organic chemicals with log K_{ow} of generally < 3.0, including nitrogen-containing pesticides, PPCPs, illicit drugs, hormones, phenoxyacids, alkylphenols benzophenone, caffeine, perfluorooctanoic acid, perfluorooctanesulfonic acid, and certain organic wastewater originated contaminants in wastewater effluents, fresh and salt water.

26) **Describe the difference between volume- and flow-proportional automated samplers.**

In the volume-proportional sampling, subsamples of a constant volume are collected at variable time intervals according to the flow. In the flow-proportional sampling, subsamples are collected at a constant time interval, but sample volumes vary according to the flow.

Problems

1) **Three wells are located in x and y plane at the following coordinates: Well 1 (0,0), Well 2 (100 m, 0), and Well 3 (100 m, 100 m). The ground surface is level and the distance from the surface to the water table in Wells 1, 2, and 3 are in the order of 10.0 m, 10.2 m, and 10.1 m, respectively. Draw the well diagram and determine the flow direction and the hydraulic gradient.**

Wells 1, 2, 3 have water levels of 10.0 m, 10.2 m, and 10.1 m, respectively, from the ground surface. Hence, Well 1 has the highest hydraulic head (H–10.0 m), whereas Well 2 has the lowest hydraulic head (H–10.2 m), where H is the elevation of ground surface above sea level.

One can sketch a schematic diagram as shown below. First, locate point D along line AB (i.e., Well 1 to Well 2, a well with the highest hydraulic head to a well with the lowest hydraulic head). Point D has the same hydraulic head as Well 3 (the well with an intermediate hydraulic head). Second, draw a line from D to C – this is the equipotential line. Third, draw a line from point B perpendicular to line CD. The groundwater flow direction shall be parallel to this line. Since Wells 1 and 3 have higher hydraulic heads than Well 2, the flow direction (arrow) is as shown in the figure, rather than the other opposite direction.

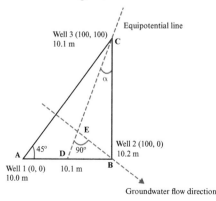

To calculate the hydraulic gradient, one needs to find the distance BE.

$\text{Tan} \, \alpha = DB/BC = 50/100 = 0.5$; hence $\alpha = 26.57°$. $EB = BC \times \sin \alpha = 100 \times \sin 26.57° = 44.7 \, \text{m}$.

$$\text{Hydraulic gradient} = \frac{\text{Head of Well C} - \text{Head of Well B}}{\text{Distance from B to the equipotential line} \, (CD)} =$$

$$\frac{(H - 10.1) - (H - 10.2)}{44.7} = 0.00224 \, \text{(dimensionless)}$$

2) **A well has a water column depth of 10 ft and diameter of 6 in. (a) What is the well volume in gallons? (b) How much water needs to be purged before samples can be collected?**

a) The volume of well (V) = volume of cylinder $= \pi r^2 h$

$$V = \pi \times \left(\frac{3}{12} \, \text{ft}\right)^2 \times 10 \, \text{ft} = 1.96 \, \text{ft}^3 \times \frac{7.48 \, \text{gallons}}{1 \, \text{ft}^3} = 14.7 \, \text{gallons}$$

Or use Eq. 5.3 directly, the constant 7.48 is the conversion factor into gallons if r and h are in the unit of foot.

b) The US EPA recommends the removal of three well-casing volumes prior to sampling. This means a total of $3 \times 14.7 = 44$ gallons of stagnant groundwater must be removed before samples can be collected.

3) **Estimate the amount (volume) of air (hence the sampling time) that should be sampled to detect toluene in workplace air using OSHA's method. The method starts with charcoal adsorption, followed by solvent desorption and subsequent analysis using GC. The recommended air volume is 12 L (0.012 m³) of air at 50 mL/min for up to 240 min. The detection limit for the overall procedure (not the GC analytical step) is 5.4 ppb$_v$ or 20.5 µg/m³ in air. This is equivalent to the minimal mass of toluene to be collected: 0.012 m³ × 20.5 µg/m³ = 0.246 µg. Toluene has a molecular mass of 92.14 g/mol.**

a) **If the workplace air contains 2 ppb$_v$ toluene, could it be detected at the recommended sampling rate and time?**

b) **If air pumping rate at 100 mL/min is also permissible, what is the minimal sampling time needed to achieve the detection limit?**

a) First, we convert 2 ppb$_v$ into µg/m³ using Eq. 2.5: $2 \times 92.14/24.5 = 7.52$ µg/m³ (MW of toluene = 92.14).

Then, we calculate the minimal mass of toluene needed based on the given detection limit of 20.5 µg/m³ if 12 L or 0.012 m³ is sampled:

$0.012 \, \text{m}^3 \times 20.5 \, \text{µg/m}^3 = 0.246 \, \text{µg}$

The minimal volume of air needed would be:

$0.246 \, \text{µg}/(7.52 \, \text{µg/m}^3) = 0.0327 \, \text{m}^3 = 32.7 \, \text{L}$

Since 32.7 L is greater than the recommended 12 L, toluene would not be detected at the suggested 50 mL/min for 240 min. In other words, toluene would be reported as "not detected" under the recommended sampling rate and sampling time.

b) If air pumping rate at 100 mL/min is permissible, what is the minimal time of sampling needed to achieve the detection limit?

$0.1 \, \text{L/min} \times t \, (\text{min}) \times 7.52 \, \text{µg/m}^3 \times (1 \, \text{m}^3/1000 \, \text{L}) > 0.246 \, \text{µg}$

Solve for t: $t > 0.246 \times 1000/(0.1 \times 7.52) = 327 \, \text{min}$

4) **The measured concentration of a PCB congener in a polyethylene passive sampler after a 30-day deployment in the field is 0.042 μg/mL. Polyethylene is a polymer with density of 0.92 g/mL. The logarithmic octanol–water partition coefficient (log K_{ow}) for PCB = 6.85. Estimate the dissolved phase concentration of this PCB congener in water.**

We first convert PCB concentration in the unit of 0.042 μg/mL to μg/g using the density of polyethylene in the sampler: $C_{PS} = (0.042\ \mu g/mL)/(0.92\ g/mL) = 0.046\ \mu g/g$.

The K_{PE-D} value can be calculated from Eq. 5.5:

$$\text{Log } K_{PS-D} = -0.59 + 1.05 \times 6.85 = 6.60$$

Thus, $K_{PS-D} = 10^{6.6} = 3.98 \times 10^6$ L/kg. We then use Eq. 5.4 to determine the PCB concentration in the dissolved phase in water:

$$C_D = \frac{C_{PS}}{K_{PS-D}} \times 1000 = \frac{0.046\ \mu g/g}{3.98 \times 10^6\ L/kg} \times \frac{1000\ g}{1\ kg} = 0.0000116\ \frac{\mu g}{L} = 0.0116\ \text{ppt}$$

5) **For Problem 4, if a grab water sample is collected and subsequently extracted into a final volume of 1 mL solvent for PCB analysis in the lab using 1 μL injection volume with GC-ECD (detection limit 0.5 μg/L). What is the minimal sample volume for a single analysis?**

Use the PCB concentration in water: 0.0000116 μg/L. If we assume a minimal volume (*V*, in liter) of water sample is collected, subsequently extracted, and concentrated to I mL solvent phase, the PCB in solvent phase must be greater than 0.5 μg/L. We then have:

$$\frac{0.0000116\ \frac{\mu g}{L} \times V(L)}{1\ mL \times \frac{1\ L}{1000\ mL}} > 0.5\ \frac{\mu g}{L}$$

Solve for *V*: *V* > 43.1 L

Hence, if we use conventional grab samples, the minimal volume of the water sample to be extracted into a final 1 mL of solution for an assumed detection limit of 0.5 μg/L for 1 μL injection volume of solvent extracted solution would be 43.1 L. For discrete sampling, this would be a very large volume of samples. Not only this is inconvenient for sample collection, transportation, and storage, but the consumption of solvent will also be very large! This exercise illustrates the importance of passive sampling for trace levels of contaminants.

Chapter 6

(Hint: Use appropriate Web sites whenever applicable)

Questions

1) **What are the hazardous characteristics described in SW-846? Which method in SW-846 is used to test toxicity characteristics of wastes?**
The hazardous characteristics described in SW-846 are: toxicity, reactivity, ignitability, and corrosivity. In SW-846, Method 1311 entitled "Toxicity Characteristic Leaching Procedure" is used to test the toxicity characteristics of waste.

2) **Which of the following is used for the acid digestion of samples containing metals: (a) EPA 3000 series, (b) EPA 3510 series, (c) EPA 3540 series, (d) EPA 3560 series?**
(a) EPA 3000 series methods include acid digestion for metals, various extraction methods for organics, and clean-up procedures for organic analysis. (b) EPA 3510 is a method entitled "Separatory Funnel Liquid-Liquid Extraction," (c) EPA 3540 is a method entitled "Soxhlet Extraction," (d) EPA 3560 is a method entitled "Supercritical Fluid Extraction of Total Recoverable Petroleum Hydrocarbons."

3) **Which of the following is related to the sample preparation and analysis of metals in the SW 846 Method: (a) Series 3000, (b) Series 6000, (c) Series 7000, (d) All of above?**
The answer is (d) all of above. (a) Series 3000 methods include acid digestion for metals, various extraction methods for organics, and clean-up procedures for organic analysis. (b) Series 6000 are the analytical methods for metals using ICP-OES, ICP-MS, XRF, isotope dilution MS, and capillary ion electrophoresis. (c) Series 7000 methods include FAAS, GFAAS, ASV, and some methods for particular elements including Hg, Cr, Se, and As.

4) **Cold vapor atomic absorption spectroscopy is commonly used for the analysis of: (a) Cr^{6+}, (b) Cr^{3+}, (c) Hg, (d) As.**
The answer is (c) Hg. In SW-846, there are currently two cold vapor atomic absorption spectroscopic methods for the analysis of mercury. Method 7470A is entitled "Mercury in Liquid Waste (Manual Cold-Vapor Technique)," and Method 7471B is entitled "Mercury in Solid or Semisolid Waste (Manual Cold-Vapor Technique)."

5) **Which of the following is used for sample preparation of volatile compounds: (a) EPA 5030, (b) EPA 3560, (c) EPA 500 series, (d) EPA 600 series?**
The answer is (a) EPA 5030, which is an SW-846 method entitled "Purge-and-Trap for Aqueous Samples." (b) EPA 3560 is a method entitled "Supercritical Fluid Extraction of Total Recoverable Petroleum Hydrocarbons." (c) EPA 500 series were published as "Methods for the Determination of Organic Compounds in Drinking Water." EPA 500 series were developed for the

Fundamentals of Environmental Sampling and Analysis, Second Edition. Chunlong Zhang.
© 2024 John Wiley & Sons, Inc. Published 2024 by John Wiley & Sons, Inc.
Companion Website: www.wiley.com/go/EnvironmentalSamplingandAnalysis2e

analysis of drinking water. (d) EPA 600 series were published as "Test Methods for Organic Chemical Analysis of Municipal and Industrial Wastewater." EPA 600 series were developed for the analysis of water and wastewater, but covers natural waters and drinking water as well.

6) **Which of the following are not directly used for the sample preparation and analysis of organic contaminants in the SW 846 Method: (a) Series 3500, (b) Series 4000, (c) Series 3600, (d) Series 8000?**

The answer is (b) Series 4000, which is a collection of immunoassays for various compounds. (a) Series 3500 are the method entitled "Organic Extraction and Sample Preparation." (c) Series 3600 are cleanup methods. (d) Series 8000 are method using GC, GC-MS, and HPLC analysis of organic contaminants, etc.

7) **Describe the difference between EPA Method 624.1 and 625.1.**

EPA method 624.1 is the analysis of volatile organic compounds (VOCs) using GC-MS, whereas EPA method 625.1 is for non-purgeable but extractable organic compounds, including mostly the semi-volatile organic compounds (SVOCs).

8) **Which EPA method would you use for the analysis of volatile solvents (e.g., trichloroethene, chloroform, toluene) in groundwater samples at μg/L concentrations?**

Method 624.1 is the analysis of volatile organic compounds (VOCs) using GC-MS.

9) **Which EPA method would you use for the analysis of industrial wastewater sludge for cadmium, mercury, and chromium to evaluate whether the sludges can be disposed in a nonhazardous waste landfill?**

SW-846. More specifically, EPA 6000 series have the ICP and ICP-MS methods and EPA 7000 series have the FAAS and GFAAS methods for most metals and particular methods for mercury and chromium analysis.

10) **Which EPA method would you use for the analysis of semivolatile and nonvolatile pesticides in wastewater from a manufacturing facility to evaluate whether the wastewater is in compliance with the facility's discharge permit?**

EPA 625 is the method for semivolatile (SVOCs) and nonvolatile (NVOCs) such as pesticides in wastewater.

11) **There is a need to determine if a manufacturing facility can landfill its solid wastes from the manufacturing of plastic resins. Compounds suspected to be in the solid wastes include semivolatile organic compounds such as 2,4,6-trichlorophenol, pentachlorophenol, pyridine, and cresols. The required EPA procedure is TCLP of the SW 846 Method 1311. You are responsible for the development of this procedure in your new lab. Explain briefly the TCLP procedure and give a list of the equipment and chemical reagents needed to develop such a method.**

The detailed procedure of TCLP can be found in Method 1311 of SW-846 online. The EPA has a list for each contaminant stating the maximum level the contaminant can reach. If the contaminant goes over the maximum level, it is classified as "Hazardous," or "Toxic." It is then assigned a Hazardous Waste Number by the EPA, and closely monitored.

Method 1311, Section 4 details apparatus and materials. Needed major items include: agitation apparatus, extraction vessels, filtration devices, filters, pH meters, zero-headspace extraction (ZHE) fluid transfer devices, analytical balance, beakers and Erlenmeyer flasks, watchglass, magnetic stirrers.

Method 1311, Section 5 details reagents for this procedure, including reagent water, HCl, HNO_3, NaOH, glacial acetic acid (CH_3CH_2COOH), extraction fluid, analytical standards, etc.

12) **Describe the difference and similarity between OSHA methods and NIOSH methods.**

Both NIOSH and OSHA are created under the Occupational Safety and Health Act of 1970. Their missions are to protect people and workers' safety and health; therefore, their methods are designed to monitor the workplace health and safety. NIOSH is a research agency under the Center for Disease Control (CDC), whereas OSHA is an agency under the Department of Labor. Because of the nature of these two agencies, OSHA methods have the enforcement authority for regulatory compliance, whereas NIOSH methods only give recommendations without any authority for enforcement.

13) **Which of the following is likely the ASTM method: (a) 239.2; (b) D3086-90, (c) TO-2, (d) 4500-Cl D?**

The answer is (b) D3086-90. All ASTM methods are named Dxxxx. (a) 239.2 appears to be an EPA method, (c) TO-2 is one of the air toxin (TO) methods for the measurement of toxic organic chemicals called hazardous air pollutants (HAPs). (d) 4500 CI D is the APHA's Standard Method for nonmetallic constituents – chloride in particular.

14) **From the online database from the National Environmental Method Index (NEMI) (http://www.nemi.gov). (a) Find a list of TOC in water methods by all agencies. (b) What is EPA Method 425.1? Download a hard copy of Method 425.1.**

a) Under "Analyte Search," type TOC. The search results from this database can be output in Excel. Selected summary outputs are given in the table below. These methods are published by the US EPA (415.1, 415.3), ASTM (D5904, D6317), APHA's Standard Methods (5310B, 5310C, 5130D), and EPA's SW-846 method (9060A). Contents of the database are updated from time to time, so students' answers might slightly vary from what are shown below.

Method ID	Method descriptive name	Method source	Source method identifier
5452	Carbon (all forms) in water	ASTM	D5904
5718	Total organic carbon by persulfate-UV or heated-persulfate oxidation	Standard methods	5310C
7614	TOC by wet oxidation	Standard methods	5310 D
9348	Total organic carbon in water and wastes by carbonaceous analyzer	EPA-OSW	9060A
5438	Carbon (all forms) in water	ASTM	D6317
5452	Carbon (all forms) in water	ASTM	D5904
5717	Total organic carbon by high-temperature combustion	Standard methods	5310B
5717	Total organic carbon by high-temperature combustion	Standard methods	5310B
7614	TOC by wet oxidation	Standard methods	5310 D
7228	Dissolved and total organic carbon and UV absorbance at 254 nm in source water and drinking water	EPA-NERL	415.3
9348	Total organic carbon in water and wastes by carbonaceous analyzer	EPA-OSW	9060A
5438	Carbon (all forms) in water	ASTM	D6317

5403	Total organic carbon	EPA-NERL	415.1
5403	Total organic carbon	EPA-NERL	415.1
5718	Total organic carbon by persulfate-UV or heated-persulfate oxidation	Standard Methods	5310C
7228	Dissolved and total organic carbon and UV absorbance at 254 nm in source water and drinking water	EPA-NERL	415.3

b) If the method number is known, we can perform the "General Search" by "Method Number" search criteria. By clicking the method number, we will be able to download the entire detailed method in pdf format. EPA method 425.1 is to determine methylene blue active substances (MBAS, i.e., anionic surfactants) in drinking water, surface water, and domestic and industrial wastes. The method is not applicable to saline waters.

15) **From the online database of the National Environmental Method Index (NEMI) (http://www.nemi.gov). (a) Find a list of immunoassay methods for organics in soil/sediment. (b) Find a list of available methods for dioxin under the National Pollutant Discharge Elimination System (NPDES).**

a) Using "General Search," select "Soils/Sediment" under "Media Name," and select "Immunoassay" under "Instrumentation." This NEMI search will give you a list of all immunoassays currently available for any organic contaminants in soil and sediment.

Method ID	Source method identifier	Method descriptive name	Media name	Method source
5626	EP 020	DDT immunoassay	Soils/sediment	Envirologix
5642	ET013	PCB (polychlorinated biphenyl) immunoassay	Soils/sediment	Envirologix
5639	7000301	PCP (pentachlorophenol) in soils/sediment by immunoassay	Soils/sediment	SDI
5635	70606	PAH in soils/sediment by immunoassay	Soils/sediment	SDI
5636	7061301	PAH in soils/sediment by immunoassay	Soils/sediment	SDI
5644	70950	Silvex (2,4,5-TP) in soils/sediment by immunoassay	Soils/sediment	SDI
5625	73100	DDT in soils/sediment by immunoassay	Soils/sediment	SDI
5619	73110	Chlordane (cyclodiene) in soils/sediment by immunoassay	Soils/sediment	SDI
5645	74200	Toxaphene in soils/sediment by immunoassay	Soils/sediment	SDI
5628	76300	Lindane in soils/sediment by immunoassay	Soils/sediment	SDI
5638	A00111/A00128	PCP (pentachlorophenol) in soils/sediment by immunoassay	Soils/sediment	SDI
5641	A00134/A00137	PCB (polychlorinated biphenyl) in soils/sediment by immunoassay	Soils/sediment	SDI

(continued)

5632	A00157/A00160	PAH in soils/sediment by immunoassay	Soils/sediment	SDI
5634	A00201/A00204	PAH in soils/sediment by immunoassay	Soils/sediment	SDI
5621	A00216/A00256	Dieldrin (cyclodiene) in soils/sediment by immunoassay	Soils/sediment	SDI

b) This can be done with "Regulatory Search" by typing "Dioxin" under the "Analyte Name" and selecting "National Pollutant Discharge Elimination System (NPDES)" under "Regulation." The search results are given in the table below.

Method ID	Method source	Source method identifier	Regulation	Regulation name	Revision information	Method descriptive name
5248	EPA-EAD	613	NPDES	National Pollutant Discharge Elimination System	40 CFR Part 136, Appendix A (Current Edition)	Qualitative confirmation and quantification of 2,3,7,8-TCDD via GC/MS after a screen using Method 625
5350	EPA-EAD	1613	NPDES	National Pollutant Discharge Elimination System	Oct-94	Dioxins and furans by HRGC/HRMS

16) **Find the approved methods for the parameters listed below by filling in the blanks. Use only the most recent revision when appropriate.**

Parameter	EPA	Standard method	ASTM	USGS/AOAC/Other
Arsenic, Total (ICP-OES)				
Cadmium, Total (ICP-MS)				
Bromide (IC)				
Toluene (GC/MS)				

See table below for appropriate methods.

Parameter	EPA	Standard method	ASTM	USGS/AOAC/Other
Arsenic, Total(ICP-OES)	200.5	3120 B-2011	D1976-12	—
Cadmium, total (ICP-MS)	200.8	3125 B-2011	D5673-16	993.14, I-4472-97
Bromide (IC)	300.1	4110-B, -C, -D-2011	D4327-17	993.30, I-2057-85
Toluene (GC/MS)	624.1, 1624B	6200 B-2011	—	O-4127-96, O-4436-16

17) **What is the primary difference between quality assurance (QA) and quality control (QC)?**

QA is more of a management system ensuring QC is working properly, and QC is a system of technical activities (such as adding blanks, spikes, duplicates, etc.) to meet certain data specifications. The use of a QA/QC program and adequate protocols will help identify and reduce errors associated with sampling, sample preservation, sample transportation, sample preparation, and sample analysis.

18) **Describe the difference between field blanks and trip blanks.**

Field blanks are designed to detect sample contamination that can occur during field operation or shipment. They are prepared in the field using clean water (for water samples) or sand (for soil/sediment samples). Trip blanks are designed only for VOCs analysis. They are prepared before going into the field by filling sample containers with clean water or sand, and then kept closed, handled, transported to the field, and then returned to the lab in the same manner as the other samples. Thus, any contamination of VOCs during shipping, handling, and analysis can be evaluated for QA/QC purposes.

19) **Describe the difference between PE/LCS and MS/MSD.**

PE/LCS stands for performance evaluation (PE) samples and laboratory control samples (LCS), respectively. PE samples are prepared by third party and submitted blind to the lab, whereas LCS are prepared by the laboratory so that the lab knows the contents of the sample. MS/MSD stands for matrix spike and matrix spike duplicates. They are designed, respectively, to test the matrix effect on the analytical accuracy and precision.

20) **Explain what type of the field QC samples should be used in the following cases:**

   a) **To know if a bailer is contaminated in the groundwater sampling.**

   b) **To know if airborne contaminants will be responsible for sample contamination during shipment from and to the lab.**

   c) **To know if an error (trace contamination) could occur during the entire sampling operation, shipment, and handling.**

   d) **To know if any sample variation exists in the field.**

   e) **To compare if the soil pollution source is due to the atmospheric deposition of an upwind smelter.**

   a) To know if a bailer is contaminated in the groundwater sampling, one would use an equipment blank as it is used to detect any contamination from a sampling equipment.

   b) To know if airborne contaminants (VOCs) will be responsible for sample contamination during shipment from and to the lab, one would use a trip blank (travel blank). Trip blanks are prepared prior to going into the field by filling in containers with clean water (for water sample) or sand (for soil samples).

   c) To know if an error (trace contamination) could occur during the entire sampling operation, shipment, and handling, one would use a field blank as it undergoes the full sampling operation, shipping, and handling process of an actual sample. It is designed to detect sample contamination that can occur during field operation and shipment.

   d) To know if there is any sample variation existing in the field, one would use field replicates (also referred to as field duplicates and split samples) as it is used to assess error associated with sample heterogeneity, sampling methodology, and analytical procedures.

   e) To compare if the soil pollution source is due to the atmospheric deposition of an upwind smelter, one would collect background samples in upstream of area(s) of contamination where there is little or no chance of migration of the contaminants of concern.

21) **Explain what type of laboratory QC samples should be used in the following cases:**

   a) **To determine the effects of matrix interferences on the analytical accuracy of a sample.**

   b) **To determine the analytical precision of a sample batch.**

c) **To establish that laboratory contamination does not cause false-positive results**.

d) **To determine whether memory effects (carryover) are present in an analytical run with an instrument**.

e) **To know whether an extraction procedure is appropriate or not**.

    a) To determine the effects of matrix interferences on analytical accuracy of a sample, one would use a spiked sample/matrix spike to evaluate the accuracy using % recovery data.

    b) To determine the analytical precision of a sample batch, one would use lab duplicates.

    c) To establish that lab contamination does not cause false positive results, one would use a method blank (reagent water) to detect any level of contamination in the lab.

    d) To determine whether memory effects (carryover) are present in an analytical run with an instrument, one would use an instrument blank with analyte-free solvent or water. This is usually injected into the instrument between high- and low-concentration samples.

    e) To know whether an extraction procedure is appropriate or not, one would use a preparation blank. The preparation blank will be extracted exactly the same way as the actual samples.

22) **The table below has a list of QC parameters and the primary means of general QC procedure. Match the right column of each corresponding QC procedure to the left column. Explain.**

QC parameter	Place a letter here	QC procedure
1. Accuracy		a. Analysis of blanks
2. Precision		b. Analysis of matrix spikes
3. Cross-contamination		c. Analysis of replicate samples
4. Extraction efficiency		d. Analysis of reference materials or samples of known concentrations

The matches below are self-explanatory. Generally, blanks are for the detection of contamination of any kind, spikes are to obtain % recovery so we will know the analytical accuracy, and replicates are for the analytical precision.

QC Parameter	Place a letter here	QC procedure
1. Accuracy	B	A. Analysis of blanks
2. Precision	C	B. Analysis of matrix spikes
3. Cross-contamination	A	C. Analysis of replicate samples
4. Extraction efficiency	D	D. Analysis of reference materials or samples of known concentration

23) **Discuss how the chemical used in a surrogate spike is different from chemical used in a matrix spike.**

Surrogate spikes use compounds that are similar to the analyte in chemical composition, extraction, and chromatography, but are not normally found in the environment or the samples to be measured. The chemical spiked during the recovery test usually is the analyte of interest. Typically, surrogate spikes are used in trace organic determination methods such as GC, GC/MS, and HPLC. They are used to access retention times (for instrument stability), percentage recovery (accuracy), and method performance.

24) **Twenty soil samples have been collected and sent to the laboratory for the analysis of lead contents. (a) What would be the appropriate analytical method? (b) What QC samples are required for this analysis?**

    a) To start with, one can use NEMI database to do an "Analyte Name" search (i.e., lead) by the "Search Criteria" (i.e., soils/sediment). This gives several relevant methods, including: NOAA's methods for marine sediment using 170.1 (ICP/MS), 140.0 (GFA), and 160.0 (XRF); USGS's I-5399 for bottom materials (FLAAS), and US EPA's 376.3 for sediment by acidification and gravimetry. It is important to note that not all these methods are appropriate, and that these are not the only methods able to analyze lead in soils. To illustrate, for example, XRF generally cannot be used for trace metal analysis, and EPA's gravimetry method obviously will not be sensitive enough for trace analysis. The remaining NOAA and USGS methods should be ok for the analyst to follow. However, if the US EPA method is preferred, then one should use the waste methods listed as Method 6000 or Method 7000 in the SW-846 online database.

    b) In this case, a group of 20 is considered as a batch. Field QC samples required for this analysis are: one equipment blank, one field blank, and one field duplicate. Lab QA/QC sample required should consider: preparation (digestion) blank, matrix spike, PE sample, and reference materials. Students' answers may vary, but blank, spike, and duplicate should be considered in both field sampling stage and lab analysis stage.

25) **Based on the information in the text and literature search, recommend the test methods by filling in the blanks with appropriate method numbers.**

Analytes	EPA method 100–600 series	SW-846 method	APHA method 23rd edition	USGS method
General physical properties of water				
Metals				
VOCs				
SVOCs				

See table below for appropriate methods.

Analytes	EPA Method 100–600 Series	SW-846 method	APHA method 23rd edition	USGS method
General physical properties of water	100, 300	—	Part 2000	P-xxxx-yr
Metals	EPA 200	3000, 6000, 7000	Part 3000	I-1472-87
VOCs	EPA 624	5000, 8000	Part 6200	O-4127-96
SVOCs	EPA 625	3500, 3600, 8000	Part 6xxx	O-5404-02

Problems

1) **How many milliliters of 5.5 M NaOH are needed to prepare 300 mL of 1.2 M NaOH?**

Equation 6.1 can be used for the dilution calculation:

$$5.5\,M \times V(mL) = 1.2\,M \times 300\ mL$$

$$V = 1.2 \times 300 / 5.5 = 65.45\ mL$$

2) **Calculate the volume of a spike stock solution in mL that should be added to dilute and prepare a 1-liter sample of 1000 ppm Cu if the desired spike concentration is 5 ppm Cu.**

$C_1 = 1000$ ppm, $C_2 = 5$ ppm, $V_1 = ?, V_2 = 1$ L

Applying Eq. 6.1: $C_1V_1 = C_2V_2$

$(1000 \text{ ppm}) \times (V_1) = (5 \text{ ppm}) \times (1L)$

$V_1 = (5\times)/(1000) = 0.005$ L $= 5$ mL stock solution.

3) **A 100-μL stock solution containing 1,000 mg/L Pb was added to a 100-mL groundwater sample as the matrix spike and acid digested. The Pb concentration measured by ICP-MS was 0.95 mg/L after a ten-fold dilution. Calculate the % recovery.**

The true concentration of the spiked sample is:

$$Pb\left(\frac{mg}{L}\right) = \frac{\text{mass of Pb in spike}}{\text{Volume of spike solution}} = \frac{100 \text{ μL} \times 1,000 \frac{mg}{L} \times \frac{1 \text{ mL}}{1,000 \text{ μL}}}{10 \text{ mL} + 100 \text{ μL}}$$

$$= \frac{0.1 \text{ mL} \times 1,000 \frac{mg}{L}}{10.1 \text{ mL}} = 9.90 \frac{mg}{L}$$

The measured concentration $= 0.95$ mg/L $\times$ dilution factor (10) $= 9.50$ mg/L

$$\%\text{Recovery} = \frac{\text{measured concentration}}{\text{True concentration}} \times 100 = \frac{9.50 \text{ mg}/L}{9.90 \text{ mg}/L} = 95.95\%$$

4) **Chromium (Cr) concentration can be determined by colorimetric spectrometry. The calibration curve was determined to be: $y = 0.045 \times C$ (μg/mL) + 0.0005, where y is the absorbance reading, and C is the concentration (μg/mL). A 10-mL surface water sample was spiked with a Cr stock solution to reach a concentration of 6 mg/L. The matrix spike has an absorbance of 0.250 against a DI water as the blank. This original surface water sample has an absorbance reading of 0.012. Calculate the % recovery.**

The net absorbance due to chromium: $A = 0.250 - 0.012 = 0.238$

Calculate the concentration (C) using calibration equation: C (μg/mL) $= (0.238 - 0.0005)/0.045$
$= 5.278$ μg/mL $= 5.278$ mg/L

% Recovery $= (5.278/6) \times 100 = 87.96\%$

5) **You are designing a sampling program and you are in charge of developing the field sampling QA/QC plan. Your crew will be on-site at an abandoned landfill for two days to collect the following samples: (a) 14 water samples from groundwater monitoring wells. The samples will be collected by pumping each well with a portable peristaltic pump and 25–50 feet of Teflon tubing into clean sampling containers. (b) 29 soil samples that will be taken using a hand auger from depths of 1–3 meters. The hand auger is essentially a rotating shovel, and the samples will be removed from the auger by scraping them into clean sampling containers.**

How many and what type of QA/QC samples are necessary? Fill in the following table by: (1) labeling the type of QA/QC samples across the top row; and (2) listing the correct number of QA/QC samples necessary for each matrix type.

Matrix	Type and number of QA/QC samples
Ground water	
Soil	

The table below only includes QA/QC samples for <u>field</u> sampling. QA/QC samples for subsequent <u>lab</u> analysis are excluded in this table. Fourteen groundwater samples can be considered as one batch, and 29 surface soil samples can be considered as one or two batches. Students' answers may vary for this question.

Matrix	Type and number of QA/QC samples				
	Equipment blank	Field blank	Trip blank	Field duplicates	Background sample
Ground water	1 per device per day	1 per day	1 per each cooler (VOCs only)	1 per matrix/10% of samples	1 per matrix
Soil	1 per device per day	1 per day	1 per each cooler (VOCs only)	1 per matrix/10% of samples	1 per matrix

6) **Benzene concentrations were detected in three monitoring wells that are expected to represent the various degrees of contamination at a contaminated site: (a) 20 μg/L, (b) 5 μg/L, and (c) 2 μg/L. A bailer was used to collect these groundwater samples, but it shows 1 μg/L of benzene in the equipment blank on this sampling event. Suggest the DQA and DUE for the result of each monitoring well based solely on this information.**

   a) The measured concentration of benzene at 20 μg/L in the monitoring well is greater than 10 times the concentration of the equipment blank. Therefore, the result is considered real and can be used directly to compare against the regulatory limit for the decision-making of site remediation.

   b) The measured concentration of benzene is greater than three times the blank, but less than 10 times the concentration of the blank. Therefore, the result is real, but it indicates quantitative uncertainty. Additional site investigations may be needed, including an evaluation of the sampling protocol.

   c) The measured concentration of benzene is less than three times the blank; thus, the presence of benzene in the groundwater sample may not be real. The data user should be warned to take an appropriate action.

7) **Toluene (an VOC) was measured by Method 8260 at a concentration of 7.5 mg/kg in a soil sample. Its regulatory limit (soil remediation standard) is 9 mg/kg. A surrogate toluene-d_8 was measured to have a recovery of 25%. Suggest the DQA and DUE for the result on this soil sample based solely on this information.**

   The recovery of 25% for VOCs is outside the range of the acceptance criteria of 70–130% for Method 8260 (Table 6.8), so the VOC results may be biased low. Since the measured concentration of 7.5 mg/kg is just below the regulatory limit of 9 mg/kg, the reported low bias indicates the results should not be used to confirm that toluene is present at a concentration less than the regulatory level. Before a firm conclusion can be made, the investigator should seek additional lines of evidence including QC data.

8) **Benzo(a)pyrene (an SVOC) was measured by Method 8270 at a concentration of 20 mg/kg in a soil sample. Its regulatory limit (soil remediation standard) is 0.2 mg/kg. The surrogate compound benzo(a)pyrene-d_{10} was measured to have a recovery of 165%. Suggest the DQA and DUE for the result on this soil sample based solely on this information.**

   The recovery of 160% for SVOCs is outside the range of the acceptance criteria of 30–130% (Table 6.8), so the SVOC results may be biased high. However, since the measured concentration of 20 mg/kg can be considered well above the regulatory limit of 0.2 mg/kg, the reported QC nonconformance should have no bearing on the usability of the results, which points to the need for the remediation of this contaminated site.

9) **A citizen's coalition is concerned about atrazine contamination in groundwater from a local pesticide manufacturer. A homogenized groundwater sample from a monitoring well was split and sent to three non-accredited research labs. The compiled results from these three labs are as follows:**

Lab	Atrazine (µg/L)	*RPD* (%)	Spike[a]	*R* (%)	Method blank
A	2.3, 3.5		1.15		1.3
B	8.2, 7.3		4.13		0.1
C	3.5, 6.7		4.52		4.2

[a] Based on a matrix spike containing 5 µg/L atrazine

a) Calculate the *RPD* (%) and *R* (%) for each lab.
b) From the performance-based analysis, which lab provides the most reliable results according to the available PARCCS?

a) The calculated *RPD* (%) and *R* (%) for each lab are shown in the table below.

Lab	Atrazine (µg/L)	*RPD* (%)	Spike	*R* (%)	Method blank
A	2.3, 3.5	41.3	1.15	23.0	1.3
B	8.2, 7.3	11.6	4.13	82.6	0.1
C	3.5, 6.7	62.8	4.52	90.4	4.2

b) Lab B is overall the most reliable because it has the high enough *R*%, lowest *RPD* (%) and lowest method blank, even though its recovery is not the highest as Lab C. Readers can find some more interesting discussions about this performance-based analysis from Cancilla DA (2001), Integration of Environmental Analytical Chemistry with Environmental Law: The Development of a Problem-Based Laboratory, *Journal of Chemical Education* 78, 12, 1652.

Chapter 7

Questions

1) **Define the following terms: aqua regia, acid bath, acetone wash, Type I water, reagent water, PPE, SDS, primary standard, Mohr pipet, mineral acidity.**

Aqua regia: It is a mixture of concentrated $HCl:HNO_3$ at a 3:1 volume ratio. This mixture can even dissolve gold and many precipitates. It is a very corrosive substance and should be used in cleaning only when required.

Acid bath: It is an acid solution containing 10–15% HNO_3 or 10–15% HCl. This acid solution (bath) is used for the cleaning of labwares (glass or plastic) for metal analysis (except for N analysis). Under acidic conditions, metals attached to container walls are dissolved or desorbed. Acid bath is not supposed to be used to soak metallic items (caps, spatulas, and brushes with metal parts).

Acetone wash: Labwares are rinsed with pure acetone. This cleaning procedure is typically employed for labwares used for all trace organic contaminant analysis.

Type I water: This is the highest quality water designated by the ASTM. It is used for tests requiring minimum interference and bias. Type I water can be prepared by distillation, deionization, or reverse osmosis followed by polishing with mixed bed deionizer and passing through a 0.2-μm pore size membrane filter to further remove trace organics, particulate matter, and bacteria.

Reagent water: This is loosely defined as the type of water without detectable concentration of compound or element to be analyzed at the detection limit of the analytical method (APHA/AWWA/WEF, 2017).

PPE: PPE stands for personal protective equipment. They are used for the prevention of hazards to the body from an environmental, chemical, respiratory, or inflammatory danger, including items like suits, safety glasses, gloves, earplugs, hard hats, safety footwear, etc.

SDS: SDS stands for safety data sheets. SDS states what the hazards are for the chemical that is being transported, and how to deal with them.

Primary standard: These are the calibrating chemicals that their accurate mass (concentration) can be determined. The exact concentration can be prepared from a reagent that is highly pure, stable, and has no waters of hydration.

Mohr pipette: A Mohr pipette has graduations that end before the tip. They differ from serological pipettes in that the serological pipettes have graduation marks continued to the tip. Both

Fundamentals of Environmental Sampling and Analysis, Second Edition. Chunlong Zhang.
© 2024 John Wiley & Sons, Inc. Published 2024 by John Wiley & Sons, Inc.
Companion Website: www.wiley.com/go/EnvironmentalSamplingandAnalysis2e

Mohr pipette and serological pipette are not as accurate as volumetric pipette (typically fixed volumes at 1, 2, 5, 10, 25, 50, 100 mL) because of the imperfectness of the internal making of the pipettes.

Mineral acidity: It is also termed methyl orange acidity because of the methyl orange used as the indicator. There are two indicators used in the titration procedure for the measurement of acidity. When NaOH is titrated, the first indicator (methyl orange) changes color that signals acidity contributed from mineral acids. When NaOH continues to be titrated, the second indicator (phenolphthalein) changes color from colorless to pink at pH 8.3. This signals the end point of pH 8.3, corresponding to the acidity from both mineral acids and weak acids (i.e., the total acidity).

2) **Illustrate how to prepare 2 N HCl and 2 N H_2SO_4 solutions from concentrated HCl and H_2SO_4.**

First, we need to know the normality of concentrated HCl and H_2SO_4. With the normality given in Table 7.2 (page 195, concentrated HCl = 11.6 N, and concentrated H_2SO_4 = 36.0 N), we can calculate the volume needed using: $N_1 \times V_1 = N_2 \times V_2$ equation, or:

$$V_2 = \frac{N_1 \times V_1}{N_2}$$

Volume of concentrated HCl needed: $V_2 = \frac{N_1 \times V_1}{N_2} = \frac{2N \times 1,000 \text{ mL}}{11.6N} = 172.41 \text{ mL}$

Volume of concentrated H_2SO_4 needed: $V_2 = \frac{N_1 \times V_1}{N_2} = \frac{2N \times 1,000 \text{ mL}}{36.0N} = 55.56 \text{ mL}$

Therefore, one needs 172.41 mL of concentrated HCl or 55.56 mL of concentrated H_2SO_4 to prepare 1.0 liter of 2N HCl or 2 N H_2SO_4. Since the addition of concentrated HCl and H_2SO_4 to water is an exothermal reaction, one needs to add the concentrated acid into water (rather than adding water into the acid) and finally dilute the solution into 1.0 liter.

3) **Explain why $K_2Cr_2O_7$ has to be used to calibrate the concentration of $Na_2S_2O_3 \cdot 5H_2O$ which is commonly used in titration.**

This is because $Na_2S_2O_3$ solution is not a stable chemical reagent; it will be readily oxidized by oxygen in the air during the storage time. In addition, the chemical used to prepare this solution is a hydrated one, i.e., $Na_2S_2O_3 \bullet 5H_2O$; its exact mass might not be accurate. On the other hand, the oxidizing agent $K_2Cr_2O_7$ is a primary standard – meaning that it is a stable compound, and its solution with an exact concentration can be prepared. Therefore, the concentration of $Na_2S_2O_3$ solution should be calibrated with the known concentration of $K_2Cr_2O_7$ solution prior to its use.

4) **What does it mean if a pipette is labeled as "5 in 1/10 ml, TD 20 °C?"**

This pipette will deliver up to 5.0 mL. It has division marks every 0.01 mL and is calibrated at 20°C. TD means the inner film will not be part of the delivered volume. However, an inner uniform liquid film should be formed upon emptying. To ensure this, the glassware must be thoroughly cleaned prior to use.

5) **Explain the operational difference for pipettes labeled as "TD" or "TC."**

TD = to deliver; TC = to contain. Pipettes designated TD will be accurate only when the inner surface is clean so that water wets it immediately and forms a uniform film upon emptying. This thin film will not be a part of the delivered volume. If the pipette is designated TC, it is calibrated that the film should be part of the contained volume. In other words, the thin film remaining inside should be flushed out with a suitable solvent. The pipettes with TC are usually used for viscous liquids.

6) **List all active and passive measures for a healthy and safe lab practice.**
The "active" measures are those "preventive" ones before the hazards take place, whereas the "passive" measures are those "protective" techniques used to passively minimize the hazard exposure after the hazards or incidents have occurred. Active measures should always be implemented first, including the knowledge of SDS, various physical/chemical/biological hazards, and their prevention. Passive measures include, but are not limited to, PPE, eye goggles, gloves, earplugs, hardhats, and the like.

7) **What action would you take in the following situation?**
   a) **Your lab mate placed broken glass pipettes (not the disposable ones) in the paper waste bin.**
   b) **Your colleague went to a vending machine downstairs while he was performing a COD reflux experiment.**
   c) **Your colleague is doing extraction using chloroform on her workbench.**
   d) **You saw a facility personnel rolling a gas cylinder while it is not capped.**
   e) **Your research assistant placed volumetric flasks in oven to dry for the experiment next day.**

   a) Broken glass pipettes should be placed in a specific waste bin for broken glass. They should not be placed in a regular trash bin.
   b) Since COD reflux has some potential safety hazards (running water, hot plates, and electricity), one should not leave this experiment unattended.
   c) Since chloroform is a toxic fume (VOC), the extraction should be performed under a well-ventilated fume hood to minimize any inhalation exposure.
   d) A gas cylinder should be capped, and it should not be rolled over on the lab floor. It should be chained and transported with a cart.
   e) Volumetric flasks should not be oven-dried because high oven temperature will likely affect its accurate volume.

8) **What is the primary difference between reflux and distillation?**
Reflux apparatus is used to accelerate a chemical reaction by maintaining a steady elevated temperature. The condenser avoids the loss of the boiling liquid as circulating water cools the vapors, which drip back into the reaction vessel. Reflux is used for chemical oxygen demand (COD) measurement. Distillation is used to purify liquids or remove solvents. It is used for the separation of the compounds dissolved in a solvent. In environmental analysis, distillation is used for the measurement of ammonium, CN^-, sulfide, and phenol.

9) **What is the major difference between liquid–liquid separatory funnel extraction and Soxhlet extraction in terms of the applicable type of sample matrices?**
Liquid–liquid (L-L) extraction uses two immiscible liquids (solvents) in a separatory funnel to extract analyte of interest. It is used to extract analyte from one liquid (usually water) into another liquid (usually organic solvent). Soxhlet is a special extraction device used for solid samples such as soil, sediment, sludge, or fish tissue. Solid samples are held in a thimble that allows solvent to repeatedly extract analyte in the sample during repeated extraction–condensation–evaporation–extraction cycles.

10) **What oven temperature is used to dry samples for suspended solid measurement? Why is this constant temperature important? Why are different temperatures of 103 °C, 180 °C, and 550 °C used in solid measurement?**
About 104±1°C is the temperature used to dry total suspended solids, whereas 180±2°C is used to dry total dissolved solids. About 550±50°C is used to dry volatile solids. "Ash" is the

residue at 550±50°C in a muffle furnace. The temperature at which the residue is dried has an important bearing on results, because weight losses due to volatilization of organic matter, mechanically occluded water, water crystallization, and gases from heat-induced chemical decomposition, as well as weight gains due to oxidation, depend on temperature and time of heating.

11) **Explain why only a single titration step is needed for alkalinity measurement if the water pH is lower than approximately 8.3.**

When HCl is used to titrate the sample, the first indicator, phenolphthalein, is used to signal a pH change from higher than 8.3 to a lower pH, which measures the phenolphthalein alkalinity. If the water sample has an initial pH lower than 8.3, then the phenolphthalein indicator will not change its color (i.e., phenolphthalein alkalinity is zero). The color of water samples will only change when pH drops below around 4.5 (corresponding to methyl orange indicator). This measures the total alkalinity (i.e., the methyl orange alkalinity).

12) **Explain why acidity is not present if a water sample has a pH greater than approximately 8.5. What is the major difference between pH and acidity?**

Just like the absence of alkalinity if a water sample is very acidic (low pH), the acidity will be zero if the water is very alkaline (high pH). However, pH and acidity are two different parameters. pH is an intensity parameter, which is independent of sample volume (e.g., the solution pH of 0.01 N HCl will be the same regardless of the solution volume), but acidity is a capacity parameter, which has the unit of mg equivalent of $CaCO_3$ per liter of water (mg/L as $CaCO_3$).

13) **Discuss the major species responsible for the acidity in natural water and possible contributing species in polluted water.**

Species responsible for acidity in *natural water* include CO_2 (from air or bacterial degradation), $H_2PO_4^-$, H_2S, protein, fatty acids, and salt of trivalent metals such as hydrated Fe^{3+} and Al^{3+}. In *polluted water*, acidity may be caused by free mineral acids (H_2SO_4, HCl) from metallurgical industry, acid mine drainage, acid rain, and organic acid waste.

14) **Discuss the major species responsible for the alkalinity in natural water and possible contributing species in polluted water.**

Species responsible for alkalinity in *natural water* is mainly H_2CO_3 (i.e., CO_2 dissolved in water). Species contributing to alkalinity in *polluted* waters are NH_3 and salts of weak acids such as borate, silicate, and phosphate (i.e., the conjugate bases of HBO_3, H_4SiO_4, H_3PO_4), and salts of organic acids.

15) **Are Na^+ and K^+ included in the hardness calculation? Why or why not?**

Water hardness is caused by multivalent metallic cations primarily the divalent Ca^{2+} and Mg^{2+} in the natural water. They form chemical precipitates with anions such as CO_3^{2-} and SO_4^{2-}. These precipitates will clog water pipes or form scale, thereby reducing heat transfer efficiency in water heaters and boilers. Anions and monovalent cations such as Na^+ and K^+ do not contribute to hardness.

16) **Write all chemical reactions involved in the measurement of DO using Winkler (iodometric) method.**

Fixing of dissolved oxygen: $Mn^{2+} + 2OH^- + \frac{1}{2}\,O_2 \rightarrow MnO_2\,(s) + H_2O$

Acidification and liberation of I_2: $MnO_2\,(s) + 2I^- \rightarrow Mn^{2+} + I_2 + 2H_2O$

Titration with starch as the indicator: $I_2 + 2S_2O_3^{2-} \rightarrow S_4O_6^{2-} + 2I^-$

Add all the above three reactions: $2S_2O_3^{2-} + 2H^+ + \frac{1}{2}\,O_2 \rightarrow S_4O_6^{2-} + S_4O_6^{2-} + H_2O$

From this overall reaction, we can see that ½ mole of dissolved oxygen will consume 2 moles of titrate (sodium thiosulfate). The ratio of O_2 to $S_2O_3^{2-} = \frac{1}{2}: 2 = 1 : 4$.

17) **Write the chemical reactions for the measurement of: (a) COD, (b) chloride using Mohr method, (c) residual chlorine using iodometric back titration, (d) cyanide.**

a) COD

The first step is dichromate digestion, where organic matter is oxidized into CO_2 and water, organic nitrogen in a reduced state is converted to NH_4^+:

$$C_nH_aO_bN_c + dCr_2O_7^{2-} + (8d + c)\,H^+ \rightarrow nCO_2 + (a + 8d - 3c)/2\,H_2O + cNH_4^+ + 2dCr^{3+}$$

where $d = 2n/3 + a/6 - b/3 - c/2$.

After dichromate digestion is complete, the excess of dichromate is titrated with ferrous ammonium sulfate $(Fe(NH_4)_2(SO_4)_2)$:

$$6Fe^{2+} + Cr_2O_7^{2-}\,(\text{excess}) + 14H^+ \rightarrow 6Fe^{3+} + 2Cr^{3+} + 7H_2O$$

The endpoint of the above titration is indicated by a chelating agent 1,10-phenanthroline (ferroin). When all $Cr_2O_7^{2-}$ is reduced, ferrous ions (Fe^{2+}) react with ferroin to form a red-colored complex:

$$Fe^{2+} + 3C_{12}H_8N_2 \rightarrow Fe\{C_{12}H_8N_2\}_3\,(\text{red})$$

Note that $Cr_2O_7^{2-}$ has a yellow to orange brown color depending on the concentration, and Cr^{3+} has a blue to green color. So the color of the solution during the titration starts with an orange brown, and then a sharp change from blue-green to reddish brown, which corresponds to the color of $Cr_2O_7^{2-}$, Cr^{3+}, and $Fe\{C_{12}H_8N_2\}_3$, respectively.

b) Chloride using Mohr method

$AgNO_3$ is used to titrate chloride (Cl^-): $AgNO_3 + Cl^- \rightleftarrows AgCl + NO_3^-$.
The endpoint is reached when excess Ag^+ reacts with potassium chromate (CrO_4^{2-}) as an indicator to form an orange Ag_2CrO_4: $2Ag^+ + CrO_4^{2-} \leftrightarrows Ag_2CrO_4$ (orange).

c) Residual chlorine using iodometric back titration

Free chlorine reacts with KI to liberate I_2: $Cl_2 + 2KI \rightarrow I_2 + 2KCl$.
The liberated I_2 is immediately reacted with an excess amount of $Na_2S_2O_3$:

$$I_2 + 2Na_2S_2O_3\,(\text{excess}) \rightarrow 2NaI + Na_2S_4O_6$$

The remaining $Na_2S_2O_3$ is back titrated with I_2 as the titrant (the iodometric back titration method) instead of using $Na_2S_2O_3$ as the titrant (the iodometric titration method).

d) Cyanide

Hydrogen cyanide (HCN) is liberated from an acidified sample by distillation and purging with air through a scrubbing solution containing NaOH. Cyanide concentration in the scrubbing solution is then titrated with standard silver nitrate $(AgNO_3)$ to form the soluble cyanide complex, $Ag(CN)_2^-$. As soon as all CN^- has been complexed and a small excess of Ag^+ has been added, the excess Ag^+ combines with a silver-sensitive rhodamine indicator (*p*-dimethyl-aminobenzal-rhodamine, $C_{12}H_{12}N_2OS_2$), which immediately turns solution from yellow to brownish-pink. The titration is based on the following reaction:

$$2CN^- + Ag^+ \rightarrow \left[Ag(CN)_2\right]^-$$

18) **During the reflux process of COD measurement, if the solution turns into green color, what causes this and what is the solution to this problem?**

After a water sample is mixed with sulfuric acid and oxidizing agent $(K_2Cr_2O_7)$, the solution should remain the color of $K_2Cr_2O_7$. A green color indicates that the sample has a very high COD value such that all hexavalent chromium $(K_2Cr_2O_7)$ has been reduced into trivalent from (Cr^{3+}) in the solution. Trivalent chromium has a light blue to green color depending on the concentration. If this occurs, the reflux procedure should not proceed. A remedy for this problem is to use a smaller volume of sample, and if this is not possible, then add a predetermined excess amount of K_2CrO_7.

19) **Typically, wastewater samples have COD > BOD$_5$. Explain if it is possible for a wastewater sample to have much larger value of COD than BOD$_5$ (COD $\gg$ BOD$_5$). Discuss the cause of why or why not.**

When a water sample contains a lot of nonbiodegradable compounds, then it is likely COD $\gg$ BOD$_5$. Nonbiodegradable organic compounds cannot be metabolized by bacteria in the BOD$_5$ procedure, but can be oxidized chemically through oxidizing agents used in COD measurement, such as K$_2$Cr$_2$O$_7$.

20) **Define the difference between chloride, free chlorines, and combined chlorines. How do they differ in their disinfection capacity.**

The chloride ion (Cl$^-$) is formed when the element chlorine (Cl) picks up one electron to form an anion (negatively charged ion) Cl$^-$. It is not effective in disinfection at all.

HOCl and OCl$^-$ are termed as "free chlorine residuals." These are the two species in maintaining the disinfection capacity in killing bacteria and pathogens. HClO is a weak acid; it is about 100 times more effective than OCl$^-$.

The chloroamines are called "combined chlorine residuals." They are less effective in disinfection but can last longer than HOCl and OCl$^-$. Therefore, chloroamines must be maintained in the water distribution systems before the drinking water reaches the consumer.

21) **Explain what types of interference (positive or negative) will be caused in the presence of a reducing agent or oxidizing agent for DO measurement.**

Oxidizing agents liberate iodine (I$_2$) from iodides (I$^-$). Since iodine is an oxidizing agent, the presence of oxidizing agents will consume more Na$_2$S$_2$O$_3$ during the titration. This will result in an overestimate of the DO value (i.e., positive interference). On the other hand, some reducing agents will reduce iodine to iodide, resulting in an underestimated DO value (i.e., negative interference).

22) **List three substances that interfere with the Winkler method, and indicate which modification would be used to overcome each interference.**

NO$_2^-$, Fe^{2+}, and suspended solids are the three common interferences during DO measurement. The modified Winkler methods to address these three interferences are the use of NaN$_3$, KMnO$_4$, and alum flocculation, respectively. The azide (NaN$_3$) modification effectively removes interference caused by nitrite, which is the most common interference in biologically treated effluents and incubated BOD samples. The permanganate modification is used if ferrous iron (Fe^{2+}) is present. When the sample contains 5 mg/L or more ferric iron salts, potassium fluoride (KF) is used as the first reagent in the azide modification or after the permanganate treatment for ferrous iron. Alternately, Fe(III) interference can be eliminated by using 85–87% phosphoric acid (H$_3$PO$_4$) instead of sulfuric acid (H$_2$SO$_4$) for acidification. This procedure has not been tested for Fe(III) concentrations above 20 mg/L. The alum flocculation modification is used if suspended solids are present; or the copper sulfate-sulfamic acid flocculation modification is used for activated-sludge mixed liquor.

23) **Describe the difference between BOD and COD in terms of the chemicals each method represents.**

BOD measures biological oxygen demand, i.e., the oxygen needed by aerobic bacteria to degrade organic substances in a water sample in a five-day period (BOD$_5$). COD measures the oxygen demand under chemical oxidizing conditions (e.g., K$_2$CrO$_7$). COD can't tell the difference between biologically oxidizable and biologically inert organics. BOD relates COD to indicate water quality, i.e., a water sample with a high COD value usually is also high in BOD. However, since K$_2$CrO$_7$ is a strong oxidizing agent, it will also oxidize some chemicals that are not likely to be degraded in nature by aerobic bacteria. To this end, COD is typically higher than BOD.

24) **Explain why (a) diluted water, (b) bacterial seed, and (c) initial oxygen saturation are all needed in the BOD measurement.**

BOD is a common biological parameter that measures the contents of organic pollutants in wastewater. Since this is an aerobic bacteria-based procedure, it requires the presence of bacteria to start with (inoculation of bacteria seed) and the presence of oxygen throughout the duration of the BOD measurement (diluted water and initial oxygen saturation). In addition, it needs nutrients (Ca, Mg, Fe salts and phosphate) and the absence of toxic chemicals for a BOD to be accurately measured.

25) **Explain the difference between direct iodometric titration and iodometric back titration. How is the color change different at the end point of titration?**

Free chlorine and chloramines stoichiometrically liberate iodine (I_2) from KI at pH of 4 or less. With iodometric titration method, $Na_2S_2O_3$ is used to titrate the liberated I_2. Iodine (I_2) is reduced back to I^-. Blue color (the product between I_2 and starch as an indicator) disappears at the end point.

With iodometric back titration method, the liberated I_2 is immediately reacted with an excess amount $Na_2S_2O_3$:

$$I_2 + 2Na_2S_2O_3 \left(\text{excess} \right) \rightarrow 2NaI + Na_2S_4O_6$$

The remaining $Na_2S_2O_3$ is back titrated with I_2 as the titrant instead of using $Na_2S_2O_3$ as the titrant (the iodometric titration method). When the endpoint is reached (i.e., all excess $Na_2S_2O_3$ is consumed), I_2 reacts with indicator starch; the solution turns blue.

26) **Describe the sample pretreatment method for sulfide measurement using titration procedure.**

Sulfide samples are preserved by the addition of zinc acetate and NaOH to maintain pH > 9.0. The precipitate of zinc sulfide (ZnS) is then oxidized to sulfur with the addition of excess iodine in an acid solution. The remaining iodine is determined by titration with sodium thiosulfate until the blue iodine–starch complex goes away.

27) **Explain the following: (a) why acid is added for oil and grease measurement; (b) why cyanide is added for hardness measurement; (c) why zinc acetate is used for sulfide measurement; (d) why NaOH is used during the distillation of cyanide.**

a) Why acid is added for oil and grease measurement?

The aggregate water/wastewater parameter "oil and grease" is measured with an acidification and hexane extraction followed by a gravimetric measurement. Fatty acids form a precipitate as Ca or Mg salts with soaps. This makes them insoluble in organic solvents. With the acidification using HCl (pH ~1.0), fatty acids will be released as free fatty acids:

$$Ca \left(C_{17}H_{35}COO \right)_2 + 2H^+ \rightarrow 2C_{17}H_{35}COOH + Ca^{2+}$$

The free fatty acids can be extracted by hexane. The extracted phase (oil and grease) is drained to a tared distilling flask through anhydrous Na_2SO_4 to remove the water. Hexane is removed by evaporation in a distillation device (at 85°C), and the tared flask is weighed a second time and the gain in weight of the flask is "oil and grease."

b) Hardness is the measure of primary Ca^{2+} and Mg^{2+} in a water sample. When titration method is used, EDTA (ethlenediaminetraacetic acid) combines with Ca^{2+} and Mg^{2+} as well as other metals that will consume EDTA and thereby overestimating hardness. The addition of cyanide (CN^-) will eliminate such interference because CN^- is a strong complexing agent.

c) Under an acidic condition, sulfides of various forms (termed acid-soluble sulfide) can be released as a gaseous H_2S. Zinc acetate [$Zn(CH_3COOZn)_2$] is added in a gas scrubbing solution, which will combine H_2S to form ZnS precipitate. Sulfide in the form of ZnS can be measured subsequently.

d) Cyanides in water samples are usually present in the form of metal complexes. These metal complexes are first converted into gaseous hydrogen cyanide (HCN) under an acidic condition. HCN from this acidified sample is collected by distillation and purging with air through a scrubbing solution containing NaOH, where NaOH reacts with HCN forming NaCN. The scrubbing solution containing NaCN is then measured by volumetric titration.

Problems

1) **Calculate the volume of concentrated H_2SO_4 (18 M) required to make 1 L of 0.5 N H_2SO_4.**

   A solution of 18 M H_2SO_4 is equivalent to 36 N H_2SO_4 because each mole of H_2SO_4 can dissociate 2 moles of H^+.

   Applying Eq. 6.1: $36\ N \times V = 0.5\ N \times 1\ L$

   $V = 0.0278\ L = 27.78\ mL$

2) **Calculate the volume of concentrated $HClO_4$ (71%, density = 1.67 g/mL) required to make 500 mL of 0.2 N $HClO_4$ (MW = 100.5).**

   Applying Eq. 7.1 to calculate the molarity of concentrated $HClO_4$,

   $$\text{Molarity}\,(M) = \frac{10 \times \text{Percentage}\,(\%) \times \text{Density}\,(g/mL)}{MW\,(g/mL)} = \frac{10 \times 71 \times 1.67}{100.5} = 11.8\ M$$

   For $HClO_4$ with one dissociated H, its molarity is the same as normality. Applying Eq. 6.1 ($N_1 V_1 = N_2 V_2$):

   $11.8N \times V = 500\ mL \times 0.2\ N$

   $V = 8.47\ mL$

3) **Without using Eq. 7.1, for a concentrated HNO_3 solution (MW = 63.01 g/mol) with an active ingredient of 71% and density of 1.42 g/mL, verify its molar concentration is 16.0 M.**

   The key is to realize that the % of active acid ingredients is defined as the weight %, for example, 71% means 71 g acid/100 g solution. Now, we can use density to convert the mass of acid per unit volume of solution:

   $$\frac{71\ g\ HNO_3}{100\ g\ Solution} \times \frac{1.42\ g\ Solution}{1\ mL\ Solution} \times \frac{1000\ mL\ Solution}{1\ L\ Solution} = \frac{71 \times 1.42 \times 1000}{100} \times \frac{g\ HNO_3}{1\ L\ Solution}$$

   $$= 10 \times 71 \times 1.42 \times \frac{g\ HNO_3}{1\ L\ Solution}$$

   Now, we divide this by the molecular weight to obtain mol/L or M:

   $$\frac{10 \times 71 \times 1.42 \times \dfrac{g\ HNO_3}{1\ L\ Solution}}{\dfrac{63.01\ g\ HNO_3}{1\ mol\ HNO_3}} = \frac{1071 \times 1.42}{63.01} \times \frac{mol\ HNO_3}{1\ L\ Solution} = \frac{71 \times 1.42 \times 1000}{63.01}\ M$$

   We now derive Eq. 7.1 for the conversion from % and density to M:

   $$\text{Molarity}\,(M) = \frac{10 \times \text{Percentage}\,(\%) \times \text{Density}\,(g/mol)}{MW\,(g/mol)}$$

4) **A sludge sample containing 35% moisture (wet basis) has a pyrene concentration of 7.35 mg/kg (dry basis). Report the pyrene concentration in sludge on wet basis.**

Rearranging Eq 7.3, we have:

$$\text{Concentration (wet basis)} = \text{Concentration (dry basis)} \times \frac{(100 - \% \text{ moisture})}{100}$$

$$= 7.35 \frac{mg}{kg} \times \frac{100 - 35}{100} = 4.78 \frac{mg}{kg}$$

5) **A sludge sample was collected after sludge was dewatered from a belt-dewatering equipment; the moisture content was 12%. This sludge sample was further air-dried and measured for its copper content at 80 mg/kg (dry basis). What is the copper concentration in mg/kg on a wet basis?**

$$\text{Concentration (wet basis)} = \text{Concentration (dry basis)} \times \frac{(100 - \% \text{ moisture})}{100}$$

$$= 80 \frac{mg}{kg} \times \frac{100 - 12}{100} = 70.4 \frac{mg}{kg}$$

6) **A 100-mL water sample was used for suspended solid measurement. The mass of weighting bottle plus filter paper was 25.6257 g before filtration and 25.6505 g after titration. What is the TSS in mg/L?**

Mass of total suspended solids = 25.6505 g – 25.6257 g = 0.0248 g = 24.8 mg

TSS (mg/L) = 24.8 mg/100 mL = 24.8 mg/0.1 L = 248 mg/L

7) **A wastewater sample is measured for its solids species using the Standard Method 2540. The results from the direct measurements are: TS = 755 mg/L, TVS = 323 mg/L, TSS = 264 mg/L, and VSS = 87 mg/L. Determine TDS, VDS, FDS, and TFS.**

Applying 7.4a: TS = TSS + TDS, 755 = 264 + TDS, **TDS** = 755–264 = 491 mg/L

Applying 7.4b: TS = TVS + TFS, 755 = 323 + TFS, **TFS** = 755–323 = 432 mg/L

Applying 7.4c: TSS = VSS + FSS, 264 = 87 + FSS, FSS = 264–87 = 177 mg/L

Applying 7.4f: TFS = FSS + FDS, 432 = 177 + FDS, **FDS** = 432–177 = 255 mg/L

Applying 7.4d: TDS = VDS + FDS, 491 = VDS + 255, **VDS** = 491 – 255 =236 mg/L

8) **A water sample from a public swimming pool is collected for total alkalinity measurement. A 50-mL pool water consumed 3.46 mL of 0.02 N HCl to reach the endpoint pH of 4.3. What is the total alkalinity in mg/L as $CaCO_3$? Is this pool water ok for swimming?**

$$\text{Alkalinity (mg/L as } CaCO_3) = \frac{N \times V \times 5 \times 10^4}{V_S} = \frac{0.02 \times 3.46 \times 5 \times 10^4}{50} = 69.2 \text{ mg/L}$$

An alkalinity of 69.2 is below the minimum of 80. This pool water is acidic and may be irritating to skin. The low alkalinity will also make the pH swing a lot due to the low buffering capacity.

9) **If 3.0 mL of 0.02 N H_2SO_4 is required to titrate 200 mL of sample to the phenolphthalein end point, what is the phenolthalein alkalinity as mg/L of $CaCO_3$? If additional 20.0 mL of H_2SO_4 is needed to reach the methyl orange point, what is the total alkalinity as mg/L of $CaCO_3$?**

Equation 7.5 is the formula for the calculation of acidity; the same formula can be used to calculate alkalinity, although the two-step titrations are exactly the opposite.

$$\text{Phenolphthalein alkalinity} = \frac{N \times V \times 5 \times 10^4}{V_s} = \frac{0.02 \times 3.0 \times 5 \times 10^4}{200} = 15.0 \frac{mg}{L} \text{ as } CaCO_3$$

$$\text{Total alkalinity} = \frac{N \times V \times 5 \times 10^4}{V_s} = \frac{0.02 \times (3.0 + 20.0) \times 5 \times 10^4}{200} = 115 \frac{mg}{L} \text{ as } CaCO_3$$

10) **A water sample has Ca^{2+} and Mg^{2+} concentration of 25.0 mg/L and 17.0 mg/L. Calculate the hardness in mg $CaCO_3$/L (atomic weight: Ca = 40, Mg = 24).**

$$\text{Hardness} = \sum M^{2+} \left(\frac{mg}{L}\right) \times \frac{\text{Equivalent weight of } CaCO_3}{\text{Equivalent weight of } M^{2+}}$$

Equivalent weight of $CaCO_3 = 100/2 = 50$, $Ca^{2+} = 40/2 = 20$, $Mg^{2+} = 24.4/2 = 12.2$

$$\text{Hardness} = 25.0 \times \frac{50}{20} + 17.0 \times \frac{50}{12.2} = 132.2 \frac{mg}{L} \text{ as } CaCO_3$$

11) **A 25-mL groundwater sample was added to 25-mL DI water for a two-fold dilution. The volume of 0.015 mol/L EDTA to reach the titration end point was 6.28 mL. Calculate the water hardness in mmol/L and mg $CaCO_3$/L.**
   - The titration reaction can be written as: $M^{2+} + Na_2EDTA \rightarrow 2Na^+ + M\text{-}EDTA$
   - The above reaction shows: moles of metal $(Ca^{2+} + Mg^{2+})$ = moles of EDTA
   - The moles of EDTA = molar concentration of EDTA × volume of EDTA = 0.015 mol/L × 0.00628 L = 9.42×10^{-5} moles
   - The molar concentration of Ca^{2+} and $Mg^{2+} = 9.42 \times 10^{-5}$ moles/25 mL = 9.42×10^{-2} mmol/0.025 L = **3.768 mmol/L**
   - Hardness expressed in mg/L as $CaCO_3 = 3.768$ mmol/L × (100 mg $CaCO_3$/1 mmol $CaCO_3$) = **376.8 mg/L as $CaCO_3$** (The molecular weight of $CaCO_3 = 100$)
   - Note that the information on "two-fold dilution with DI water" is disregarded, because sample volume before dilution (25 mL) was used in the calculation, and the dilution with de-ionized water does not change the hardness.

12) **To prepare EDTA solution for hardness measurement, 8.54 g EDTA disodium salt ($C_{10}H_{14}N_2O_8Na_2 \cdot 2H_2O$, MW = 372.2) was dissolved in 1000 mL DI water. After calibration, its concentration was measured to be 22.82 mmol/L. A 50-mL water sample consumed 6.06 mL EDTA solution. (a) What is the actual concentration of EDTA in mmol/L compared to the measured concentration of 22.82 mmol/L? (b) What is the hardness of this water sample in mg $CaCO_3$/L?**

a) What is the theoretical (actual) concentration of EDTA in mmol/L?

$$EDTA \left(\frac{mmol}{L}\right) = \frac{8.54 \text{ g}}{L} \times \frac{1 \text{ mmol}}{372.2 \text{ mg}} \times \frac{1{,}000 \text{ mg}}{1g} = 22.95 \frac{mmol}{L}$$

This theoretical concentration (22.95 mmol/L) compares favorably to the measured (calibrated) concentration of 22.82 mmol/L.

b) What is the hardness of this water sample in mg $CaCO_3$/L?

$$\text{Hardness} \left(\frac{mmol}{L} \text{ as } CaCO_3\right) = \frac{22.82 \frac{mmol}{L} \times 6.06 \text{ mL} \times \frac{1 \text{ L}}{1{,}000 \text{ mL}}}{50 \text{ mL} \times \frac{1 \text{ L}}{1{,}000 \text{ mL}}} = \frac{0.138 \text{ mmol}}{0.05 \text{ L}} = 2.77 \frac{mmol}{L}$$

$$\text{Hardness} \left(\frac{mg}{L} \text{ as } CaCO_3\right) = 2.77 \frac{mmol}{L} \times \frac{100 \text{ mg } CaCO_3}{1 \text{ mmol } CaCO_3} = 277 \frac{mg}{L} \text{ as } CaCO_3$$

13) **(a) Explain why the conversion factor in Eq. 7.7 is 8000. (b) If a 200-mL sample was used for dissolved oxygen (DO) measurement, it requires 8.60 mL of 0.0275 N $Na_2S_2O_3$ solution. What is the DO in mg/L?**

a) The formula to calculate DO is: $DO \left(\frac{mg}{L}\right) = \frac{N \times V \times 8{,}000}{V_s}$

where N and V are the normality and volume (mL) of $Na_2S_2O_3$, respectively, V_s is the sample volume (typically 200 mL is withdrawn from 300 mL sample).

The overall reaction of the Winkler method is:

$$2S_2O_3^{2-} + 6H^+ + \tfrac{1}{2}\,O_2 \rightarrow S_4O_6^{2-} + 3H_2O$$

From $S_2O_3^{2-}$ to $S_4O_6^{2-}$, the oxidation number of S changes from +2 to +2.5. Since the net change is +1 per molecule, meaning one mole of electron per mole $S_2O_3^{2-}$ is lost to O_2. This also means that 1 mol of thiosulfate ($S_2O_3^{2-}$) is equal to 1 equivalent of the same species. $N{\times}V$ equivalents of $S_2O_3^{2-}$ will be equal to $N{\times}V$ moles of $S_2O_3^{2-}$ or $N{\times}V/4$ mol of O_2. Converting this number of moles into mass of O_2 in mg, this equals to $(N{\times}V/4){\times}32{\times}1000$ or $N{\times}V{\times}8{,}000$, where 32 is the molecular weight of O_2.

b) $N = 0.0275$ N, $V = 8.60$ mL, $V_s = 200$ mL

$$\mathrm{DO}\left(\frac{\mathrm{mg}}{\mathrm{L}}\right) = \frac{N \times V \times 8{,}000}{V_s} = \frac{0.0275 \times 8.60 \times 8{,}000}{200} = 9.46\,\frac{\mathrm{mg}}{\mathrm{L}}$$

14) **A standard procedure by APHA's method was followed to determine DO in water. The volume of 0.025 M $Na_2S_2O_3$ consumed was 6.8 mL in titrating 200-mL solution. What is the DO in mg/L?**

The valence change of S from $S_2O_3^{2-}$ to $S_4O_6^{2-}$ is +2 to +2.5. The net change is therefore +1 per molecule of $S_2O_3^{2-}$. Hence, the molecular weight of $S_2O_3^{2-}$ is equal to the equivalent weight of $S_2O_3^{2-}$. A solution of 0.025 M $S_2O_3^{2-}$ will also have its concentration of 0.025 N. In addition, $V= 6.80$ mL, $V_s = 200$ mL.

$$\mathrm{DO}\left(\frac{\mathrm{mg}}{\mathrm{L}}\right) = \frac{N \times V \times 8{,}000}{V_s} = \frac{0.025 \times 6.80 \times 8{,}000}{200} = 6.80\,\frac{\mathrm{mg}}{\mathrm{L}}$$

The above calculation shows that if the concentration of $S_2O_3^{2-}$ is exactly equal to 0.025 N, then the numerical value of DO (in mg/L) will be equal to the volume of $S_2O_3^{2-}$ consumed during titration.

15) **Glucose is oxidized according to: $C_6H_{12}O_6 + 6O_2 \rightarrow 6CO_2 + 6H_2O$. It can be used as a standard of COD because its theoretical COD can be calculated from its known concentration. The oxidation–reduction reaction involved in COD measurement is:**

$$6Fe^{2+} + Cr_2O_7^{2-} + 14H^+ \rightarrow 6Fe^{3+} + 2Cr^{3+} + 7H_2O$$

a) **What is the theoretical COD value of a solution containing 0.50 g/L glucose ($C_6H_{12}O_6$; MW = 180)? (*Hint*: 1 mole glucose consumes 6 moles of O_2.)**
b) **Why is the equivalent weight of $K_2Cr_2O_7$ (MW=294.2) 294.2/6 = 49.03?**
c) **To prepare 500 mL 1.0 N $K_2Cr_2O_7$, how many grams of $K_2Cr_2O_7$ are needed?**
d) **List three major sources of errors in COD measurement. How might they be eliminated?**

a) From the reaction stoichiometry, the mass ratio $C_6H_{12}O_6/6O_2 = 180\colon 6{\times}32 = 180\colon192 = 1\colon1.067$.

$$C_6H_{12}O_6 + 6O_2 \rightarrow 6CO_2 + 6H_2O$$

The COD is then: 0.50 g/L $\times 1.067 = 0.533$ g/L $= 533$ mg/L.

b) The oxidation reduction reaction involved in COD measurement is:

$$6Fe^{2+} + Cr_2O_7^{2-} + 14H^+ \rightarrow 6Fe^{3+} + 2Cr^{3+} + 7H_2O$$

The above equation indicates that 6 moles of electrons are gained for every mole of $Cr_2O_7^{2-}$. The equivalent weight is therefore 1/6 of its molecular weight, i.e., $294.2/6 = 49.03$ g/equivalent.

1.0 N $K_2Cr_2O_7$ is the same as $1/6 \times 1.0 = 0.167$ M $= 0.167$ mol/L

$$0.167\,\frac{\mathrm{mol}}{\mathrm{L}} \times \frac{294.2\ \mathrm{g}}{1\ \mathrm{mol}} = 49.0333\,\frac{\mathrm{g}}{\mathrm{L}} = 24.5167\,\frac{\mathrm{g}}{500\ \mathrm{mL}}$$

c) COD measurements are prone to a variety of errors; some may be due to analysts' operational errors, and others may be related to the lack of precaution needed in dealing with samples of certain specific characteristics. For example, one does not need to add H_2SO_4 very accurately, but the addition of K_2CrO_7 must be added accurately using a volumetric pipette. Sample volume is also critical for good accuracy. A sample with a high COD (e.g., wastewater) needs a small volume of sample, whereas a very low COD sample (natural water) needs a large volume. Samples containing volatile compounds (such as volatile fatty acids from anaerobic wastewater treatment systems) should be refluxed in a closed system instead of an open reflux one. Samples of high salinity (Cl^-) such as contaminated estuarine waters should be refluxed with the addition of mercury chloride to remove the interference from chloride.

16) **Potassium hydrogen phthalate (abbreviated as PHK, $HOOCC_6H_4COOK$, MW = 204.23) is always used as the COD standard for its measurement. To prepare 1 liter of a PHK solution containing 500 mg/L COD, how many grams of PHK need to weigh? The oxidation of PHK is assumed: $2HOOCC_6H_4COOK + 15O_2 \rightarrow K_2O + 16CO_2 + 5H_2O$. (*Hints*: 2 mole PHK consumes 15 moles of O_2.)**

From the given oxidation reaction, the mass ratio of PHK to O_2 is: $2 \times 204.23 : 15 \times 32 = 408.46 : 480 = 0.8510 : 1$. A solution of 500 mg/L COD will be equivalent to $500 \times 0.8510 = 425.48$ mg/L $= 0.4255$ g/L of PHK. Therefore, 0.4255 g of PHK should be weighted using an analytical balance and then dissolved in 1 liter of DI water in a volumetric flask for a COD standard containing 500 mg/L.

17) **The table below shows the experimental data obtained during BOD_5 measurement. If data is valid for wastewater samples labeled as Seed, A, B, and C, determine the respective BOD_5 values.**

Bottle #	Sample	Sample volume (mL)	Seed volume (mL)	DO_0 (mg/L)	DO_5 (mg/L)
1	Seed	12	0.00	8.96	5.74
2	A	10	0.00	8.88	3.95
3	B	200	3.00	8.88	5.63
4	C	250	3.00	8.73	0.81

Bottle #1 (Seed Correction Factor): $(8.96 - 5.74)/12 = 0.27$ mg/L per mL
Bottle #1 (BOD_5 in seed): $(8.96 - 5.74)/(12/300) = 80.5$ mg/L
Sample A in Bottle #2 (BOD_5 without the use of microbial seed): $(8.68 - 3.95)/(10/300) = 152.7$ mg/L
Sample B in Bottle #3 (BOD_5 with the use of microbial seed): $[(8.88 - 5.63) - (0.27 \times 3.00)]/(200/300) = 3.66$ mg/L
Sample C in Bottle #4 (BOD_5 with the use of microbial seed): Results from Bottle # 4 are not used in reporting because it has an invalid final DO of less than 1.0 mg/L. The analyst should re-do the measurement by reducing the sample volume for a larger dilution ratio.

18) **A municipal wastewater sample contains 50 mg/L of ammonia (NH_3), what is the equivalent concentration of NH_3-N (i.e., NH_3 as nitrogen) in mg/L?**

$$NH_3 \text{ as nitrogen} \left(\frac{mg}{L}\right) = 50\frac{mg}{L} \ NH_3 \times \frac{14 \ mg \ N}{17 \ mg \ NH_3} = 41.2\frac{mg}{L} \ NH_3-N$$

19) **The NPDES permit limit for PO_4^{3-}–P is 1 mg/L, what is the equivalent concentration for PO_4^{3-} in mg/L?**

$$PO_4^{3-} \left(\frac{mg}{L}\right) = 1\frac{mg}{L} \ PO_4^{3-} \times \frac{(31+4 \times 16) \ mg \ PO_4^{3-}}{31 \ mg \ P} = 3.06\frac{mg}{L} \ PO_4^{3-}$$

Chapter 8

Questions

1) **Explain why SPME and headspace are nonexhaustive extraction methods?**

    In both SPME and static headspace, a small volume of the extracting phase relative to the sample volume is employed. Thus, the capacity of the extraction phase is smaller and is usually insufficient to remove most of the analytes from the sample matrix. Therefore, SPME and headspace are nonexhaustive extractions.

2) **Explain briefly: (a) why is HNO_3 rather than other acids most commonly used for acid digestion; (b) why HF digestion is used for samples containing silicates; (c) why or why not acid-preserving samples are appropriate for dissolved metals analysis.**

    a) HNO_3 acts as both an acid and an oxidizing agent in the digestion process. As a strong acid, it dissolves inorganic oxides into solutions. As an oxidizing agent/acid combo, HNO_3 can oxidize zero-valance inorganic metals and nonmetals into an ionic form. Another important factor is that metal nitrates are all water soluble. Additional advantages of using HNO_3 are: HNO_3 is an acceptable matrix for both flame and electrothermal atomic absorption; HNO_3 is also a preferred matrix for ICP-MS analysis.

    b) No acids other than HF will liberate the metal of interest from silica matrix.

    c) Dissolved metals are chemically defined as hydrated ions, inorganic/organic complexes, and colloidal dispersions. Operationally, dissolved metals are defined as the metals in an unacidified sample that pass through a 0.45-μm membrane filter. For dissolved metal analysis, samples should be filtered immediately on-site.

3) **Explain: (a) why is HCl weaker than HNO_3 when used in acid digestion; (b) why should dryness be avoided in acid digestion and why is dryness particularly harmful if $HClO_4$ is used; (c) why should $HClO_4$ not be used alone for acid digestion.**

    a) HCl is not as effective as HNO_3 in releasing metals from sample matrices. Unlike HCl, HNO_3 acts as both an acid and an oxidizing agent in the digestion process. As a strong acid, it dissolves inorganic oxides into solutions. As an oxidizing agent/acid combo, it can oxidize zero-valance inorganic metals and nonmetals into an ionic form.

    b) Dryness should be avoided in acid digestion because volatile metals such as Hg, Pb, Ca, Cd, As, Sb, Cr, and Cu are subject to loss if a digestion solution becomes dry.

    c) Digestion reaction may be violent if $HClO_4$ is used alone in digesting samples with high organic contents. Pretreat with HNO_3 before adding $HClO_4$ to oxidize most of the organic matter.

Fundamentals of Environmental Sampling and Analysis, Second Edition. Chunlong Zhang.
© 2024 John Wiley & Sons, Inc. Published 2024 by John Wiley & Sons, Inc.
Companion Website: www.wiley.com/go/EnvironmentalSamplingandAnalysis2e

4) **Describe the difference in the acid preservation procedure between total metals and dissolved metals.**

Acid preservation (HNO_3 to pH <2.0) is required in the field for total metal analysis, whereas filtration is performed prior to acid preservation for dissolved and suspended metals.

5) **Explain why speciation analysis is particularly important for Cr, Hg, As, and Se. What are the major valances (oxidation number) for each of these four elements? Describe their impact with regard to plants, and animals/humans.**

Speciation analysis is particularly important for Cr, Hg, As, and Se because various species of each of these four elements have considerably different toxicity toward plant and animal/human. Chromium has two primary valence states. The hexavalent chromium [Cr(VI)], existed as CrO_4^{2-} and $Cr_2O_7^{2-}$, is toxic to organisms (pants, animal, human), and even carcinogenic to humans. The trivalent chromium [Cr(III)], existed as Cr^{3+}, $Cr(OH)_2^+$, $Cr(OH)_2^+$, and $Cr(OH)_4^-$, is considered nonessential for plants, but an essential trace element for animals and humans.

All forms of mercury are toxic (neurotoxic) to animals and humans, including elemental mercury (Hg^0), inorganic mercury (Hg^{2+}), methyl mercury (MeHg) and other organic mercury species. The most toxic form of mercury is MeHg.

Arsenic and selenium are nonessential for plants but are essential to several animal species (As) and essential to most animals (Se). Arsenic, in its inorganic form, has two oxidation states, i.e., arsenate, designated as As(V), and the trivalent arsenite, designated as As(III). Arsenite is many times more toxic than arsenate. The predominant As(V) species in water is $H_2AsO_4^-$ at pH 3–7 and $HAsO_4^{2-}$ at pH 7–11. Under reducing conditions, the major As(III) species are $HAsO_2$ (aq) or H_3AsO_3. Organic As can also arise from industrial discharges, pesticides, and biological action of inorganic As. Unpolluted fresh water normally does not contain organic arsenic compounds, but may contain inorganic arsenate and arsenite.

Inorganic selenium (Se) exists predominately as selenate ion (SeO_4^{2-}), designated as Se(VI), and selenite ion (SeO_3^{2-}), designated as Se(IV). Other common aqueous species include Se^{2-}, HSe^-, and Se^0. Se derived from microbial degradation of seleniferous organic matter includes selenite, selenate, and the volatile organic compounds dimethylselenide and dimethyldiselenide. Nonvolatile organic selenium compounds may be released into water by microbial processes.

6) **For each one of the elements Cr, Hg, Se, and As, identify the most toxic speciation.**

Cr: Hexavalent Cr compounds, Cr(VI), have been shown to be carcinogenic by inhalation and are corrosive to tissue. Trivalent chromium, Cr(III), is considered nontoxic. Chromium is considered nonessential for plants, but an essential trace element for animals.

Hg: Mercury (Hg) is a nonessential element for both animals and humans and all forms of inorganic and organic Hg are considered to be toxic. The most notorious is, perhaps, the methylmercury because of its high toxicity and bioaccumulation in the food chain in aquatic systems.

Se and As: Both arsenic (As) and selenium (Se) are nonessential for plants but As is essential for several animal species and Se is essential to most animals. The reduced form of arsenic, arsenite, designated as As (III), is many times more toxic than arsenate, designated as As (V). Selenite, Se(IV), is many times more toxic than selenate, (Se(VI).

7) **Name five operationally defined species of heavy metals in soil using the sequential extraction method.**

Table 8.2 is a list of five fractions of metal species following a classical sequential extraction method by Tessier et al. (1979). In order of the sequential extraction, these fractions are exchangeable, acid soluble (bound to carbonates), reducible (bound to Fe and Mn oxides), oxidizable (bound to organic materials or sulfides), and residual metals.

8) **Describe/define briefly: (a) dissolved metals; (b) alkaline digestion for Cr(VI); (c) selenate ion vs selenite ion; (d) arsenate and arsenite.**

   a) Dissolved metals, defined chemically, include hydrated ions, inorganic/organic complexes, and colloidal dispersions, which are equivalent to operationally defined dissolved metals in an unacidified sample that passes through a 0.45-μm membrane filter.

   b) Alkaline digestion procedure for Cr(VI) is distinctly different from the acid digestion used for other metals. With this method, an alkaline solution containing 0.28 M Na_2CO_3/0.5 M NaOH is mixed with sample and heated at 90–95°C for 60 minutes. This treatment extracts/dissolves the Cr(VI) from soluble (e.g., $K_2Cr_2O_7$), adsorbed, and precipitated forms (e.g., $PbCrO_4$) of Cr compounds in soils, sludges, sediments, and similar waste materials. The pH of the digestate must be carefully adjusted during the digestion to maintain the integrity of the Cr species (i.e., avoid reduction of Cr(VI) or the oxidation of Cr(III)).

   c) Inorganic selenium (Se) exists predominately as selenate ion (SeO_4^{2-}), designated as Se(VI), and selenite ion (SeO_3^{2-}), designated as Se (IV).

9) **Explain: (a) why a chemical with Henry's law constant (H) of 10^{-4} atm·m³/mol cannot be separated using separate funnel L-L extraction; (b) why a solution containing a chemical with very high vapor pressure could be low in volatility.**

   a) A chemical with Henry's law constant of 10^{-4} atm-m³/mol cannot be separated using separate funnel L-L extraction because it is a volatile compound. L-L extraction can only be used for semivolatile or nonvolatile compounds.

   b) The vapor pressure of a chemical (liquid or solid) is the equilibrium pressure of a vapor above its pure liquid (or solid, but not its dissolved form in water solution). That is, the pressure of the vapor resulting from evaporation of a liquid (or solid) above a sample of the liquid (or solid) in a closed container. Henry's law constant is the ratio of the vapor phase concentration (or vapor pressure, etc.) to the liquid phase concentration when air-liquid partitioning reaches equilibrium. Normally a compound with a high vapor pressure will have a high Henry's law constant, but this correlation is not always true. A solution containing a chemical with a very high vapor pressure could be low in volatility because a compound with a high vapor pressure such as ethanol could have very low volatility if ethanol is present in water, but it evaporates quickly when a drop of pure ethanol is spilled on a table.

10) **Describe how distribution coefficient (K) and the phase ratio affect the extraction efficiency in the L-L extraction.**

    The partition coefficient of an analyte is defined as: $K = C_s/C_w$, where C_s and C_w are the equilibrium concentration in solvent and water, respectively. The extraction efficiency is as follows:

    $$E = \frac{C_s V_s}{C_s V_s + C_w V_w} = \frac{1}{1 + \frac{V_w}{V_s}\frac{1}{K}}$$

    From the above equation, the extraction efficiency (E) is independent of the initial analyte concentration, but is a function of the partition coefficient and water-to-solvent ratio (V_w/V_s). The extraction efficiency increases as partition coefficient (K) increases and decreases as the water-to-solvent ratio (V_w/V_s) increases.

11) **Describe the operational step in using Soxhlet extraction.**

    1) A dry solid sample is placed into a permeable cellulose thimble.
    2) The thimble is loaded into the Soxhlet extractor.
    3) Extraction solvent is placed in the round-bottomed flask and heated to boiling.
    4) The vapors rise through the outer chamber into the condenser and then condense and drip down onto the extracting sample.
    5) When the extraction chamber is almost full, it is emptied by a siphoning effect.
    6) The flushed solvent returns to the flask and condenses.

7) The extraction is repeated with fresh solvent that is redistilled from the solution in the bottom flask. It repeats many times (h) as necessary.

8) Extraction is complete when the solution in the chamber is the same color as the pure solvent indicating that nothing more is being extracted.

12) **Explain: (a) why acetone or acetonitrile cannot be used in solvent extraction? (b) In extracting organics from aqueous solutions using a separatory funnel, which layer (top or bottom) should be collected if benzene is used? What if chloroform is used?**

a) Acetone and acetonitrile, and a number of other solvents, are miscible in water. Thus, there will be no separate phases formed, and so no extraction can occur. Solvent extraction should be done to produce two distinct phases.

b) When using a separatory funnel, the solvent and water relationship will depend on density differences in the two liquids. For benzene (density 0.87 g/cm^3 at 25°C) and water (density 0.997 g/cm^3 at 25°C), the organics will be extracted into the benzene layer, which is the top layer. For chloroform (density 1.46 g/cm^3 at 25°C) and water, organics will be extracted into chloroform, which is the bottom layer.

13) **Explain: (a) why solid phase extraction (SPE) can only be used for liquid samples; (d) why anhydrous sodium sulfate (Na_2SO_4) is added during solid sample extraction; (b) why SPME can be used for both liquid and solid samples?**

a) A SPE (solid phase extraction) device is typically a cartridge or disk packed with silica particles bonded with organic coating. Samples must be in their liquid form, so when samples are loaded, they will flow through the cartridge and chemicals to be extracted will be partitioned into the organic coating.

b) Anhydrous sodium sulfate (Na_2SO_4) is used in the extraction for the removal of water since it is an inert drying agent for organic solutions.

c) An SPME (solid phase microextraction) uses a fiber that has a typical dimension of 1 cm $\times$ 110 µm, which is normally bonded to a stainless-steel plunger and installed into a microsyringe. This fiber can be inserted directly into liquid samples, or headspace for solid phase samples. Chemicals sorbed into the fiber coating are then desorbed by heating into a capillary GC column for subsequent analysis.

14) **Discuss the advantages and disadvantages of the following sample preparation: (a) conventional Soxhlet extraction; (b) ultrasonic extraction; (c) solid phase extraction; (d) supercritical fluid extraction.**

a) There are several advantages of Soxhlet extraction. First, it is a very thorough extraction method, and can thus be used as a benchmark to validate any new extraction method that is devised. It can also be used for any solid samples, though its main environmental application is the extraction of SVOCs from solid samples such as soil, sludge, and solid waste. It is also relatively inexpensive and simple to perform. Its disadvantages are that it is a time-consuming procedure and requires the use of large quantities of solvent.

b) Ultrasonic extraction takes relatively little time (few minutes), but it is not effective when high extraction efficiencies of analytes at very low concentrations are necessary. Also, there are concerns that ultrasonic energy may break down some organo-phosphorous compounds. The usefulness of ultrasonic extraction for a specific application (compound-matrix) should be validated with the benchmark Soxhlet extraction.

c) SPE uses a smaller volume of solvent than conventional extraction, no emulsions are produced, and the process can be readily automated, thus saving time, money, and labor. Its disadvantages are that it has a relatively higher cost than conventional extraction, and the automation that makes this an attractive method may not be available in lower-end labs.

d) SFE takes advantage of some of the unusual properties of super critical fluids (SCFs). This method can readily penetrate porous and fibrous solids and thus has a high extraction efficiency. In addition, SFE is an automated procedure that may cut analysis times down significantly. On the downside, SFE is a very costly procedure, and is extremely complicated to perform. Because of these factors, SFE will most likely be found only in the most well-equipped labs where cost is of less concern than it is in academic labs.

15) **Name several sample extraction methods that can be considered green methods.**
Compared to the traditional sample preparation using LLE and Soxhlet, green sample preparation techniques use one of the following approaches:
- Use of environmentally friendly solvents: Supercritical carbon dioxide, subcritical water, ionic liquids (ILs).
- Use of solventless techniques, such as SPE, SPME, SBSE, MSPE.
- Use of virtually solvent-free techniques, such as single-drop microextraction (SDME), liquid phase microextraction (LPME), and membrane techniques.
- Use of assisted solvent extraction through various types of energies such as microwave-assisted extraction (MAE, either pressurized or atmospheric), ultrasound-assisted extraction (UAE), and pressurized liquid extraction (PLE).
- Use of scaled-down sample preparation device such as micro total analysis system (μTAS).

16) **Discuss the advantage of: (a) microwave-assisted acid digestion over hotplate digestion; (b) Soxtec over conventional Soxhlet; (c) pressured fluid extraction over extraction with ambient temperature and pressure.**
a) Hotplate digestion is conducted under a ventilation hood specially designed for minimal exposure to metals while allowing rinse of the corrosive acids from the hood after the digestion is completed. The microwave-assisted acid digestion system uses specially designed microwave that is acid-proof along with safety features for acid fume collection and programmable temperature and pressure control. Both methods use mineral/oxidizing acids and an external heat source to decompose matrix and liberate metals in an analyzable form. The microwave-assisted acid digestion method can offer faster and more reproducible results than the conventional hot plate methods. If closed vessels are used, loss of volatile elements (Hg, Pb, As, Sb, Cr, Cu, Cd, Ca) can also be avoided.
b) Soxhlet is a rugged and well-established technique, but the major disadvantages are its time-consuming procedure (6–48 h) and a large solvent usage, which often requires a concentration step to evaporate solvent. In Soxtec, extraction is faster because of the vigorous contact between the boiling solvent and the sample. Since the concentration step is integrated into Soxtec, the extract is ready for cleanup and analysis.
c) The mechanisms for the enhanced extraction efficiency at high temperature and pressure in the PFE system compared to extractions at or near room temperature and atmospheric pressure are suggested to be the result of the increased analyte solubility and mass transfer effects, and the disruption of surface equilibria. At a high temperature, the solubility and diffusion rate of analytes are increased whereas solvent viscosity is decreased. High pressure keeps solvent liquefied above its boiling point and allows solvent to penetrate matrix readily. Elevated temperature and pressure also disrupt the analyte–matrix bonding such as van der Waal's forces, hydrogen bonding, and dipole attractions of the solute molecules and active sites on the matrix.

17) **Why does CO_2 in its supercritical state increase the extraction efficiency? What are the major mechanisms for the enhanced extraction efficiency? Why is a minor amount of solvent used in the SFE system?**

CO$_2$ in its supercritical fluid state has a gas-like property with a high mass transfer coefficient and liquid-like solvent property. It also has low viscosity and almost zero surface tension. The high diffusivity allows it to readily penetrate porous and fibrous solids. As a result, CO$_2$ in its super critical fluid state offers a good extraction efficiency. A minor amount of solvent is used only to keep the analyte in the solution after supercritical CO$_2$ is depressurized so it can be removed through the evolution of gaseous CO$_2$.

18) **Draw a schematic diagram for the supercritical extraction system (SFE).**

Figure 8.13 is a schematic diagram of an SFE device. It consists of a tank of the mobile phase (CO$_2$), a pump to pressurize CO$_2$, an oven containing the extraction vessel, a restrictor to maintain a high pressure in the extraction line, and a trapping vessel. Analytes are trapped by letting the solute-containing supercritical fluid decompress into an empty vial, through a solvent, or onto a solid sorbent material. After depressurization, CO$_2$ evolves as a gas.

19) **Explain how ionic liquids can be used in sample extraction.**

Ionic liquids are salt compounds with ionic nature. However, because their melting points are below 100°C, these salts are present as liquid below 100°C. They have unique properties including negligible vapor pressure, good thermal stability, and miscibility with water and organic solvents as well as good extractability for various organic compounds and metal ions.

20) **Draw a schematic diagram for the purge-and-trap system with GC as the analytical instrument.**

Figure 8.14a is a schematic diagram of purge-and-trap (P&T) with GC as the analytical instrument. In P&T, separation is carried out by purging the sample, sometimes done while heating, with an inert gas such as He or N$_2$ and trapping the volatile materials in a sorbent column (Tenax, silica, charcoal). The trap is designed for rapid heating so that it can be desorbed directly into a GC column. The P&T is applicable for compounds having boiling points below 200°C and are insoluble or slightly soluble in water. The purge efficiency can be improved for water soluble analytes, e.g., ketones and alcohols, when purging at an elevated temperature of 80°C as compared to 20–40 °C. P&T is often coupled with GC and programmed to operate first in the purge mode (10–12 min) and then in desorbing mode (1–2 min).

21) **Discuss the principles and compound applicability of the following sample clean-up procedures: (a) alumina; (b) Florisil; (c) silica gel.**

a) Alumina: It is a highly porous granular aluminum oxide that can adsorb interfering compounds. It allows the pH adjustment to a specific range (acid, base, neutral) and thus can be used to separate analytes from interfering compounds of different polarities. For example, EPA 3610 is a method for the analysis of phthalate esters (neutral alumina) and nitrosamines (basic alumina), and EPA 3611 is a method for the separation of petroleum waste into aliphatic, aromatic, and polar fractions (neutral alumina).

b) Florisil: It is an activated form of magnesium silicate with basic properties. It is more acidic and milder (with regard to compound decomposition) than alumina and silica gel. It has been used for cleaning pesticide residues and other chlorinated hydrocarbons, for separating nitrogen-containing compounds from hydrocarbons, and for separating aromatic compounds from aliphatic–aromatic mixtures. It is also good for separations with steroids, esters, ketones, glycerides, alkaloids, and some carbohydrates. EPA 3620 is a method for the analysis of phthalate esters, nitrosamines, organochlorine pesticides, PCBs, chlorinated hydrocarbons, and organophosphorus pesticides.

c) Silica gel: It is a regenerative adsorbent of silica with weakly acidic amorphous silicon oxide. Silica gel is the precipitated form of silicic acid (H_2SiO_3), formed from the addition of H_2SO_4 to sodium silicate. It forms very strong hydrogen bonds to polar materials and can lead to analyte decomposition. It is somewhat soluble in methanol, which should never be used as an elution solvent. Silica gel can also remove slightly polar substances from hexane solutions and serves to isolate the strictly petroleum hydrocarbon analytes. EPA 3630 is a method for the analysis of PAHs, PCBs, and derivatized phenol.

22) **Derivatization is sometimes used prior to HPLC and GC analysis: (a) what are the purposes of derivatization; (b) what types of compounds (e.g., with certain functional groups) normally need to be derivatized?**

Derivatization in chemical analysis is used to: (a) increase the volatility and decrease the polarity of compounds; (b) increase thermal stability to protect analyte from thermal degradation; (c) increase detector response by incorporating functional groups which leads to higher detector signals, e.g., CF_3 groups for electron capture detectors; (d) improve separation and reduce tailing. Common derivatization methods can be classified into four groups depending on the type of reaction applied: silylation, acylation, alkylation, and esterification.

23) **Describe: (a) silylation; (b) acylation; (c) esterification.**

a) Silylation – A derivatization method that involves replacing active hydrogen with a tri-methylsilyl (TMS) group. This procedure produces compounds that are more volatile and more thermally stable. The ease of reactivities of the functional group toward silylation follows the order: alcohol > phenol > carboxyl > amine > amide > hydroxyl.

b) Acylation – A derivatization method that adds the acyl group (RCO–) to compounds. This reduces the polarity of multifunctional compounds, such as carbohydrates and amino acids, and converts compounds with active hydrogen into esters, thioesters, and amides. It also adds halogenated functionalities to enable electron capturing ability with GC-ECD detection.

c) Esterification – A derivatization method in which an acid is reacted with an alcohol in the presence of a catalyst to form an ester, a type of compound that has a lower boiling point than the original compound.

24) **What physicochemical properties of glyphosate make it very challenging to analyze?**

The analysis of glyphosate is quite challenging due to its ionic character, low volatility, low mass, and lack of functional groups (e.g., chromophores, fluorophores) that detectors will respond to. Thus, derivatization is generally required for the analysis of glyphosate using all analytical methods.

25) **Describe the following and indicate in general what types of contaminants (i.e., with regard to solubility, volatility, etc.) are applicable: (a) dynamic headspace extraction (purge-and-trap); (b) static headspace extraction; (c) azeotropic distillation; (d) vacuum distillation.**

a) The purge-and-trap is applicable for compounds having boiling points below 200°C and are insoluble or slightly soluble in water. Problems are often encountered with the low purging efficiency of methyl *t*-butyl ether and related fuel oxygenated compounds.

b) The static headspace extraction is applicable only to compounds of high volatility from water. This requires a higher air–water partitioning coefficient (K) that can result in a high vapor phase concentration to be detected. A low headspace-to-sample volume ratio will increase concentrations of volatile analytes in the gas phase and therefore lead to a better sensitivity. Compounds with a very low volatility cannot be analyzed by static headspace analysis.

c) The azeotropic distillation is designed for nonpurgeable, water soluble, and volatile organic compounds. These are mostly small molecules including alcohol, aldehyde, and ketone as specified in EPA Method 5031. Azeotropic distillation is a nonconventional distillation technique.

d) Vacuum distillation can be used to separate organic compounds that have a boiling point below 180°C and are insoluble or slightly soluble in water (such as BTEX compounds). EPA method 5032 is based on a vacuum distillation and cryogenic trapping procedure followed by gas chromatography/mass spectrometry.

26) **Equations 8.3a and 8.9 are used to describe the liquid–liquid extraction and vapor–liquid equilibrium, respectively. Illustrate why the similarity between these two formulas.**

Equation 8.3a describes the residual concentration of analyte in the <u>aqueous phase</u> (C_w^1) after a single extraction:

$$C_w^1 = C_w^0 \frac{V_w}{KV_s + V_w} = C_w^0 \frac{1}{K\dfrac{V_s}{V_w} + 1}$$

where K is the equilibrium partition coefficient of an analyte between solvent and water sample ($K = C_s/C_w$, C_s and C_w are the equilibrium concentration in solvent and water, respectively), and V_s/V_w is the volume ratio of solvent and water sample.

Equation 8.3a can be used to calculate the equilibrium concentration of chemicals in the <u>solvent phase</u> after a single extraction:

$$C_s^1 = C_w^1 \times K = C_w^0 \frac{1}{K\dfrac{V_s}{V_w} + 1} \times K = \frac{C_w^0 K}{K\dfrac{V_s}{V_w} + 1} = \frac{C_w^0}{\dfrac{V_s}{V_w} + \dfrac{1}{K}}$$

If we use β to denote the phase ratio ($\beta = Vs/V_g$), then the above equation becomes:

$$C_s^1 = \frac{C_w^0}{\dfrac{V_s}{V_w} + \dfrac{1}{K}} = \frac{C_w^0}{\beta + \dfrac{1}{K}}$$

This equation is the same when Eq. 8.9 is rearranged, except the notations are different (i.e., K vs H, subscript "s" denotes solvent in Eq. 8.9, whereas the same subscript in Eq. 8.9 denotes aqueous sample). Note also that in liquid–liquid extraction, we are extracting chemical from initial water into a solvent, whereas in static headspace extraction, we are "extracting" chemical from initial water phase into gas phase.

Recall that Eq. 8.9 in its rearranged form below describes the final concentration of volatile compounds (C_g) in the headspace of sample vials at the equilibrium state:

$$C_g = \frac{C_o}{\beta + 1/H}$$

where C_g is the concentration of volatile analytes in the gas phase and C_o is the original concentration of volatile analytes in the (aqueous) sample, β is the phase ratio ($\beta = V_s/V_g$, where V_s and V_g are the volumes of (aqueous) sample phase and vapor phase), and H is the equilibrium partitioning coefficient between gas phase and the aqueous sample ($H = C_g/C_s$).

27) **Which of the following chemical(s) need azotropic distillation for its separation from an aqueous sample: (a) hydrophobic pesticide, (b) acetone, (c) acetonitrile, (d) methyl isobutyl ketone, (d) pyrene.**

From Table 8.6, it is known that the separation of chemical mixtures that behave like a single substance needs azotropic distillation. Acetone, acetonitrile, and methyl isobutyl ketone are all nonpurgeable, water soluble, and volatile organic compounds that cannot be separated by any conventional distillation.

28) **Name some common chemicals that are nonpurgeable VOCs.**

Many VOCs are not purgeable; hence, the purge-and-trap method commonly used for VOCs is not applicable. The azeotropic distillation is designed for these nonpurgeable, water soluble, and volatile organic compounds. These are mostly small molecules, including alcohol, aldehyde, and ketone as specified in EPA Method 5031 (see Table 8.6 for a detailed list).

Problems

1) **A clean soil sample was used as a matrix to develop an extraction method for pyrene (a hydrophobic compound), using methylene chloride (MECl) extraction. The soil has no previous contamination of this compound, so the background concentration of pyrene can be assumed zero. This analyst spiked 1.025 g soil sample with 5 mL of 10 mg/L pyrene (dissolved in acetone) stock solution, let the acetone evaporate under a hood, and then extract it into 10 mL MECl with a sonic extractor. The soil extract was then determined to contain 2.0 mg/L pyrene by HPLC method. Calculate the % recovery of the extraction. Is this an acceptable extraction method?**

$$\text{Total mass spiked in the soil} = 5\text{ mL} \times 10\frac{\text{mg}}{\text{L}} \times \frac{1\text{ L}}{1{,}000\text{ mL}} = 0.05\text{ mg}$$

$$\text{Expected concentration in soil extract} = \frac{\text{mass spiked}}{\text{volume of extract solution}}$$

$$= \frac{0.05\text{ mg}}{10\text{ mL}} \times \frac{1{,}000\text{ mL}}{1\text{ L}} = 5\frac{\text{mg}}{L}$$

Since the measured concentration is only 2 mg/L, the recovery is:

$$\text{Recovery}(\%) = \frac{\text{Measured concentration}}{\text{Expected}(\text{true})\text{ concentration}} \times 100 = \frac{2}{5} \times 100 = 40\%$$

Apparently, this sonic extraction is not an acceptable method of extraction because of the low % recovery.

2) **About 100 mL of aqueous solution containing 2,4,6-trinitrotoluene (TNT) at an initial concentration of 0.05 ppm is shaken with 10 mL hexane. After phase separation, it was determined that the organic phase has 0.4 ppm. (a) What is the partition coefficient, and what is the percent extracted? (b) If the equilibrium concentration (rather than the initial concentration in sample solution) in water phase is 0.05 ppm (this would simplify the calculation), what would then be the partition coefficient and percent extracted?**

a) If the <u>initial</u> aqueous phase concentration = 0.05 mg/L, the <u>equilibrium</u> aqueous phase concentration needs to be calculated using mass balance equation.

$$\text{Total mass of TNT} = 0.05\frac{\text{mg}}{\text{L}} \times 100\text{ mL} \times \frac{1\text{ L}}{1{,}000\text{ mL}} = 0.005\text{ mg}$$

$$\text{Mass of TNT in hexane phase at equilibrium} = 0.4\frac{\text{mg}}{\text{L}} \times 10\text{ mL} \times \frac{1\text{ L}}{1{,}000\text{ mL}} = 0.004\text{ mg}$$

$$\text{Mass of TNT in aqueous phase at equilibrium} = 0.005\text{ mg} - 0.004\text{ mg} = 0.001\text{ mg}$$

$$\text{Equilibrium concentration in water} = \frac{\text{mass}(\text{mg})}{\text{volume}(\text{L})} = \frac{0.001\text{ mg}}{100\text{ mL} \times \frac{1\text{ L}}{1{,}000\text{ mL}}} = 0.01\frac{\text{mg}}{\text{L}}$$

$$\text{Partitioning coefficient } (K) = \frac{\text{equilibrium concentration in hexane phase}}{\text{equilibrium concentration in aqueous phase}}$$

$$= \frac{0.4 \frac{mg}{L}}{0.01 \frac{mg}{L}} = 40 \, (\text{dimensionless})$$

$$\% \text{ Extracted} = \frac{\text{mass in hexane}}{\text{total mass}} \times 100 = \frac{0.004 \text{ mg}}{0.005 \text{ mg}} \times 100 = 80\%$$

b) If the <u>equilibrium</u> aqueous phase concentration $= 0.05$ mg/L, then the partitioning coefficient can be calculated directly as:

$$\text{Partitioning coefficient} (K) = \frac{\text{equilibrium concentration in hexane phase}}{\text{equilibrium concentration in aqueous phase}}$$

$$= \frac{0.4 \frac{mg}{L}}{0.05 \frac{mg}{L}} = 8 \, (\text{dimensionless})$$

Total mass of TNT = mass of TNT in water at equilibrium + mass of TNT in hexane at equilibrium

$$= 0.05 \frac{mg}{L} \times 100 \text{ mL} \times \frac{1 \text{ L}}{1,000 \text{ mL}} + 0.004 \text{ mg} = 0.005 \text{ mg} + 0.004 \text{ mg} = 0.009 \text{ mg}$$

$$\% \text{ Extracted} = \frac{\text{mass in hexane}}{\text{total mass}} \times 100 = \frac{0.004 \text{ mg}}{0.009 \text{ mg}} \times 100 = 44\%$$

3) **Liquid–liquid extraction (LLE) has long been a standard method for concentrating hydrophobic organic compounds such as pesticides and polycyclic aromatic hydrocarbons (PAHs) from water. A sample of water contaminated with a pesticide is extracted with methylene chloride for determination of this pesticide.**

a) In experiments with a standard solution of a known concentration, it is found that when a 100 mL water sample is extracted with 30 mL of methylene chloride, 75% of this pesticide is removed. What is the partition coefficient (K) of this pesticide? What fraction of this pesticide would be removed if three extractions of 10 mL of methylene chloride each were done instead?

b) The extraction efficiency of LLE depends on the hydrophobicity (hence K) of the chemical to be extracted as well as the volume and number of extractions. Demonstrate this by filling in the blanks in the table below regarding the total extraction efficiency (assume water sample volume = 100 mL).

	Extraction efficiency if K (unitless) =				
	10	25	50	100	1,000
One extraction with 30 mL					
Three extractions, 10 mL each					

a) Similar to E_w (fraction in water) in Eq. 8.1a, the fraction in organic solvent phase E_o can be derived as: $E_o = \dfrac{1}{1+\dfrac{V_w}{V_s}\times\dfrac{1}{K}}$, we can derive the formula for partitioning coefficient as:

$$K = \frac{\dfrac{V_w}{V_s}}{\dfrac{1}{E_o}-1} = \frac{\dfrac{100\text{ mL}}{30\text{ mL}}}{\dfrac{1}{0.75}-1} = 10$$

If 10 mL methylene chloride is used each time, and three extractions are performed, then the total extraction efficiency should be better than one extraction using 30 mL.

$$\text{Total after 1st extraction}: E_{o1} = \frac{1}{1+\dfrac{V_w}{V_s}\times\dfrac{1}{K}} = \frac{1}{1+\dfrac{100\text{ mL}}{10\text{ mL}}\times\dfrac{1}{10}} = 0.5\,(50\%)$$

$$\text{Total after 2nd extraction}: E_{o2} = E_{o1}+\left(1-E_{o1}\right)\times E_{o1} = 0.5+\left(1-0.5\right)\times0.5$$
$$= 0.5+0.25 = 0.75\,(75\%)$$

$$\text{Total after 3rd extraction}: E_{o3} = E_{o2}+\left(1-E_{o2}\right)\times E_{o1} = 0.75+\left(1-0.75\right)\times0.5$$
$$= 0.75+0.125 = 0.875\,(87.5\%)$$

Alternatively, we can use Eq. 8.2 to derive the total extraction efficiency E after nth extractions. For $n = 3$:

$$E_{o3} = 1-\frac{C_w^n}{C_w^0} = 1-\left(\frac{1}{K\dfrac{V_o}{V_w}+1}\right)^3 = 1-\left(\frac{1}{10\dfrac{10}{100}+1}\right)^3 = 1-0.125 = 0.875\,(87.5\%)$$

b) A summary of the calculation is shown below:

K (dimensionless) =	10	25	50	100	1000
E (1 × 30 ml) (%) =	75	88.23529	93.75	96.77419	99.66777
E (3 × 10 mL) (%) =	87.5	97.66764	99.53704	99.92487	99.9999

Example calculations for $K = 50$ are as follows:

$$\text{One extraction with 30 mL}: E_{o1} = \frac{1}{1+\dfrac{V_w}{V_o}\times\dfrac{1}{K}} = \frac{1}{1+\dfrac{100\text{ mL}}{30\text{ mL}}\times\dfrac{1}{50}} = 0.9375\,(93.75\%)$$

$$\text{Three extractions with 10 mL each}: E_{o3} = 1-\frac{C_w^n}{C_w^0} = 1-\left(\frac{1}{1+K\dfrac{V_o}{V_w}}\right)^3$$

$$= 1-\left(\frac{1}{1+50\dfrac{10}{100}}\right)^3 = 0.9953704 = 99.53704\%$$

Chapter 9

Questions

1) **Describe the differences between absorption, emission, and fluorescence.**

   In spectroscopy, *absorption* of electromagnetic radiation is the way by which the energy of a photon is taken up by matter (electrons of an atom or a molecule). Thus, the electromagnetic energy is transformed into other forms of energy, including heat. Depending on the wavelength range, the absorption-based spectroscopy can be ultraviolet-visible spectroscopy, infrared spectroscopy, or X-ray absorption spectroscopy.

   The *emission* of a chemical element or chemical compound is the electromagnetic radiation emitted by the element's atoms or the compound's molecules when they are returned to a ground state.

   Fluorescence is one special type of emission of electromagnetic radiation by a substance. A substance absorbs the light of a certain wavelength and subsequently induces the emission of light with a longer wavelength (and lower energy). The energy difference between the absorbed and emitted photons is due to thermal losses.

2) **Use molecular orbital theories to explain: (a) why can hexane and water be used as a solvent for UV absorption spectrometry; (b) why CO_2 has two sorption bands in the mid-IR range; (c) why the frequency of bonds is in the increasing order of: C-C < C=C < C≡C?**

   a) Hexane and water can be used as solvents for UV absorption spectrometry because both are transparent in the UV range ($\lambda = 200$–380 nm). Hexane [$CH_3(CH_2)_4CH_3$] is a saturated hydrocarbon. In this molecule, there are 5 sigma (σ) C–C bonds, one separating each of the 6 carbons. In water molecule (H_2O), there are two sigma (σ) O-H bonds. The amount of energy required to excite an electron for electronic transition ($\sigma \rightarrow \sigma^*$ for hexane and $\sigma \rightarrow \sigma^*$ and $n \rightarrow n^*$ for water) is very large. This amount of energy requires a short wavelength (<200 nm), which is not in the regular UV range of the UV spectroscopy.

   b) Carbon dioxide has two bands in the mid-IR range (i.e., 2.5–50 μm, or 4,000–200 cm^{-1}) due to the fact that it undergoes two types of molecular vibrations, i.e., asymmetric stretching and bending. The asymmetric stretching absorbs IR at 4.26 μm and the bending absorbs at 15 μm.

   c) This is because the stronger the bond (triple is stronger than double is stronger than single), the more energy is required to excite vibrations. Since energy is directly proportional to frequency, the triple bond C≡C requires the highest energy or the highest frequency of the infrared radiation.

Fundamentals of Environmental Sampling and Analysis, Second Edition. Chunlong Zhang.
© 2024 John Wiley & Sons, Inc. Published 2024 by John Wiley & Sons, Inc.
Companion Website: www.wiley.com/go/EnvironmentalSamplingandAnalysis2e

3) **Explain (a) how two *s* orbitals form a σ bond, (b) how two *p* orbitals form a σ bond.**

   a) When two *s* atomic orbitals overlap, the sigma bond is formed. During bond formation, energy is released as these two orbitals start to overlap till a maximum stability (minimum energy) is achieved when the two nuclei are at certain distance apart.

   b) When two *p* orbitals of two atoms interact then there are two possibilities: if the p orbitals are oriented head-to-head, then again the sigma bond results.

4) **Explain in general: (a) why is a wavelength of less than approximately 180 nm hardly used in UV spectrophotometer; (b) why is IR less sensitive than UV but more useful than UV to deduce the structure of unknown chemicals; (c) why does CO_2 and water vapor always present problem when measuring atmospheric contaminants using IR?**

   a) A wavelength of less than 180 nm is hardly used in UV spectrophotometers because usually only a low-pressure hydrogen or deuterium lamp for UV (185–375 nm) and a tungsten incandescent lamp for visible measurements are used. Wavelengths lower than 180 nm need high-energy UV lamps and the spectrometer should be operated under a vacuum condition.

   b) IR is less sensitive than UV but more useful than UV to deduce the structure of unknown chemicals. For the richness of structural information, it is because IR radiations are associated with molecule's rotational and vibrational sublevels, whereas UV only interacts with certain bondings within a molecule. A simple molecule may have various vibrational and/or rotational changes under IR, thereby providing much structural information. However, IR spectroscopy is typically less sensitive because of its low energy than UV.

   c) CO_2 and water vapor always present problems when measuring atmospheric contaminants using IR because both chemicals are able to absorb IR. The presence of high concentration of CO_2 and water vapor will interfere with the measurement of trace-level atmospheric contaminants.

5) **Define/describe the following terms: (a) the difference between absorption spectroscopy and emission spectroscopy; (b) π electrons and *n* electrons; (c) sigma (σ) antibonding molecular orbital; (d) dispersive IR and nondispersive IR.**

   a) Absorption of radiation moves the atom or molecule to a higher energy level, but when the energy at the higher state returns to the ground state it is called emission. As their name implies, all absorption spectroscopic techniques are based on the "absorption" of certain radiation, whereas all emission spectroscopic techniques are based on the "emission" of certain wavelengths. In *molecular* spectroscopy, for example, UV-visible spectroscopy and IR spectroscopy are all "absorption" based, whereas fluorescence spectroscopy is "emission" based. In *atomic* spectroscopy, flame ionization and graphite furnace atomic absorption spectroscopy are "absorption" based, whereas X-ray fluorescence and ICP-OES are "emission" based.

   b) π electrons and *n* electrons: The term "*n* electrons" refers to unshared electron pairs, i.e., in the nonbonding orbital. The term "π" refers to the electrons when they form from two *p*-orbitals by a lateral (i.e., shoulder-to-shoulder) overlapping.

   c) Sigma (σ) antibonding molecular orbital can be formed by combining either 1*s* or 2*p* atomic orbitals. An antibonding orbital can be perceived as the detraction from the formation of bond between two atoms, similar to the darkness that occurs when two light waves cancel each other out or the silence that occurs when two sound waves cancel each other out. In the case of two 2*p* atomic orbitals, a σ antibonding molecular orbital is formed when two 2*p* atomic orbitals overlap end-on (i.e., head-to-head).

d) Dispersive IR spectrometers use a grating monochromator to select wavelengths and are commonly used when a single wavelength is desired. The non-dispersive infrared (NDIR) technique spectrometers use interchangeable filters to isolate a particular wavelength for the measurement of a particular application, such as various gases in air and total organic carbon (TOC) in a TOC analyzer.

6) **Which of the following electronic transition(s) is usually concerned with UV-VIS spectroscopy of organic compounds: (a) $\sigma \to \sigma^*$, (b) $n \to \pi$ and $\pi \to \pi^*$, (c) $n \to \sigma^*$?**

UV-VIS spectroscopy of organic compounds is usually concerned with electronic transitions from $n \to \pi$ or $\pi \to \pi^*$. This is because the absorption peaks for these two types of transitions fall in an experimentally convenient region of the spectrum (200–700 nm). Only when compounds have an unsaturated group can have the electronic transition to the π bonding.

7) **Draw electron configuration ($1s^2$, $2s^2$...) for (a) C and (b) O. The atomic numbers for C and O are 6 and 8, respectively.**

a) The electron configuration of carbon (C) is $1s^2 2s^2 2p^2$

b) The electron configuration of oxygen (O) is $1s^2 2s^2 2p^4$

8) **Draw the Lewis structure of the following compounds, and indicate the type of electrons (σ, π, n) in the molecules: (a) H_2O, (b) methanol (CH_3OH), (c) formaldehyde (HCHO), (d) benzene (C_6H_6).**

(a) 2 sigma bonds, 2 lone pairs (n), (b) 5 sigma bonds, 2 lone pairs (n); (c) 3 sigma bonds, one pi bond, and 2 lone pairs (n); (d) 12 sigma bonds and 3 pi bonds. (σ bond in black; π bond in gray; lone pair (n) in blue.)

(a)   (b)   (c)   (d)

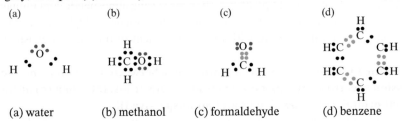

(a) water   (b) methanol   (c) formaldehyde   (d) benzene

9) **Draw the Lewis structure of the following compounds, and indicate the type of electrons (σ, π, n) in the molecules: (a) methane (CH_4), (b) ethene ($CH_2=CH_2$), (c) nitrate (NO_3^-), (d) aldehyde**

(a)   (b)   (c)   (d)

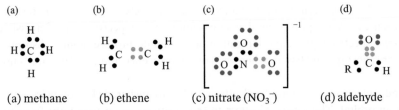

(a) methane   (b) ethene   (c) nitrate (NO_3^-)   (d) aldehyde

(a) 4 sigma bonds, (b) 5 sigma bonds, 1 pi bond, (c) 3 sigma bonds, 8 lone pairs and 1 pi bond; (d) 3 sigma bonds (for formaldehyde), 2 lone pair and 1 pi bond.

10) **Draw the electron dot formula for: (a) O_2, (b) H_2O, and (c) CH_4.**

(a) O_2 (b) H_2O (c) CH_4

$$:\ddot{O}::\ddot{O}:$$

H:Ö:

H:C:H with H above and H below

11) **The potassium permanganate solution has a purple to violet color depending on its concentrations. Which of the following range of its λ_{max} can be used for colorimetric analysis: (a) 380–420 nm, (b) 500–550 nm, (c) 550–580 nm, (d) 680–780 nm?**

According to Table 9.1, a compound with the observed "purple to violet" color (corresponding to absorbed color of "green" to "yellow-green") should have λ_{max} of 500–520 (purple) to 520–550 (violet). Thus, the answer is (b) 500–550. From literature, potassium permanganate solution has two reported λ_{max} at 526 nm and 546 nm.

12) **Rank the following compounds in order of decreasing λ_{max} :**

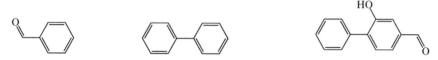

(a) benzaldehyde   (b) phenyl benzene   (c) para-aldehyde, hydroxy phenylbenzene

Compounds with a higher conjugation will have a higher λ_{max} value, thus, (c) > (b) > (a).

13) **Rank the following compounds in the order of decreasing λ_{max}:**

(a) benzaldehyde   (b) cyclohexa-2,4-dienecarbaldehyde

(c) cyclohexa-1,3-dienecarbaldehyd

A conjugated compound contains two or more double bonds alternating with single bonds. The degree of conjugation and thus the λ_{max} value for (a) is the highest with the order of (a) > (c) > (b).

14) **Explain: (a) why do you think it takes more energy (shorter wavelengths, higher frequencies) to excite the stretching vibration than the bending vibration? (Refer to Fig. 9.19 for CO_2.) (b) Why do triple C≡C bonds have higher stretching frequencies than C=C double bonds, which in turn have higher stretching frequencies than C–C single bonds?**

a) Referring to the ball-spring model of CO_2 depicted in Fig. 9.19, it is easier to bend a bond than to stretch or compress it. Therefore, stretching frequencies are higher than corresponding bending frequencies.

b) The triple bond has the three sharing pairs of electrons in the bond and therefore is stronger than a double bond which in turn is stronger than a single bond. The stronger the bond, the more energy (higher frequency) is required to excite the stretching vibration.

15) **Refer to Fig. 9.21, why the IR absorption wavenumber for C=N is greater than that for C–N?**

The double bond in C=N has the two sharing pairs of electrons and therefore is stronger than a single bond in C–N. The stronger the bond, the more energy (higher frequency, or high wavenumber) is required to excite the stretching vibration.

16) **(a) Explain what are linear molecules and nonlinear molecules. (b) Calculate how many vibrational modes are there for CO and H_2S.**

a) The difference is in the molecule structure. Linear molecules (e.g., CO_2) are in a straight line, usually caused by double- or triple-bonding scenarios. This limits the vibration modes that can occur. Nonlinear molecules (e.g., H_2O) are those that are "bent" in nature, due to single-bond physics. These nonlinear molecules tend to be easier to move, twist, etc., causing additional vibrational modes.

b) The CO molecule is a linear molecule, it has $3N-5 = 3 \times 2 - 5 = 1$ vibrational mode. H_2S is a nonlinear molecule, it has $3N-6 = 3 \times 3 - 6 = 3$ vibrational modes.

17) **Identify the correct compound that matches the IR spectrum shown below.**

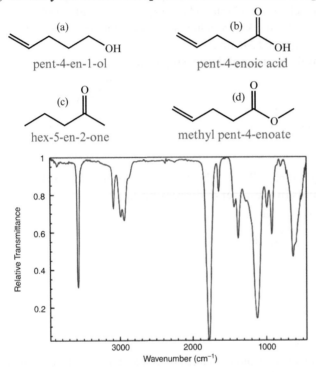

(a) pent-4-en-1-ol

(b) pent-4-enoic acid

(c) hex-5-en-2-one

(d) methyl pent-4-enoate

The answer is (b). Note that chemical (a) is also called 5-hydroxyl-1-pentene, and (b) is also called 5-pentenoic acid.

18) **Describe the difference between a typical UV spectrum and an IR spectrum.**

A spectrum in general is a plot of light intensity or power as a function of frequency or wavelength. In UV-VIS spectroscopy, a spectrum is typically a plot of absorbance vs wavelength. In infrared spectroscopy, a spectrum is typically plotted as % transmittance vs wavenumber. Since absorbance (A) is related to % transmittance (T) as: $A = -\log T$, an IR spectrum consists of inverted peaks, each representing the absorption of certain functional groups at the specific wavenumber. Note also that the x-axis (wavenumber) in the IR spectrum is not in a linear scale, and that wavenumber is related to the wavelength as follows: $\bar{\nu}\ [\text{cm}^{-1}] = \dfrac{10{,}000}{\lambda\ [\mu m]}$.

19) **Explain why the nonlinear deviation in IR is more severe than UV-VIS?**

With IR, the nonlinear deviations from Beer's law are more common than with UV and visible wavelengths. This is because infrared absorption bands are relatively narrow. For dispersive IR, the low intensity of sources and low sensitivities of detectors in this region require the use of relatively wide monochromator slit widths, leading to nonlinear relationship between absorbance and concentration.

20) **Which one of the following is true regarding Beer's law: (a) absorbance is proportional to both path length and concentration of absorbing species, (b) absorbance is proportional to the log of the concentration of absorbing species, (c) absorbance is equal to P_0/P.**

Only the statement (a) "absorbance is proportional to both path length and concentration of absorbing species" is true. This statement reflects Beer's law: $A = \varepsilon \, l \, C$, where A = absorbance, l = light path length, C = concentration, and ε = molar absorptivity (a proportionality constant).

21) **For the following compound formaldehyde:**

$$\begin{array}{c} H \\ \diagdown \\ \diagup \\ H \end{array} C = O$$

a) **Draw the electronic structure that shows all the sharing and nonsharing bonds.**

b) **Indicate the types of molecular orbitals in formaldehyde.**

a) The formaldehyde molecule should exhibit a trigonal shape overall, with a mixture of single- and double-covalent bonds. All of the bonds in formaldehyde are covalent bonds; thus, they are all sharing bonds.

b) Two types of molecular orbitals exist in formaldehyde. The first is the σ bond, which exists between the H and C atoms. This is formed by the overlap of their s atomic orbitals. The second type of bond exists between the C and O atoms. Since these atoms are double bonded, one of the two covalent bonds has already been formed as the more stable σ bond. The second covalent bond that exists between the C and O of the carbonyl group is the weaker π bond. In this case, the existence of the π bond results from the overlap between the sp^2 hybrid orbitals of the C and O atoms.

22) **Draw a schematic diagram of a: (a) UV-VIS spectrometer and (b) FTIR spectrometer.**

Students can refer to Figs 9.15 and 9.22 for UV-VIS spectrometer and FTIR spectrometer, respectively.

23) **From the following methods, do a literature/internet search (*Hint*: NEMI database and APHA's standard method book), give a list of methods that use only "UV" spectrometry (i.e., not to include visible spectrometry). (a) EPA 100–600 series methods; (b) APHA's 4000 series methods for inorganic nonmetallic constituents.**

a) Students' answers may vary. This is a good exercise for students to get familiar with the NEMI database (http://www.nemi.gov).

b) Students' answers may vary. One needs to find this handbook entitled "Standard Methods for the Examination of Water and Wastewater," 23rd Edition, edited by the American Public Health Association (APHA), American Water Works Association (AWWA) & Water Environment Federation (WEF) (2017) in a library to search for a list of these methods.

24) **Which one of the following IR instruments cannot record spectrum: (a) FTIR, (b) DIR, or (c) NDIR? Explain why?**

Nondispersive infrared (NDIR) instruments are not capable of recording an infrared spectrum. NDIR does not have the optical device to disperse infrared radiations of various wavelengths. Instead, it uses interchangeable filters to isolate a particular wavelength for measurement.

25) **Explain (a) why can *in situ* atmospheric CO₂ be monitored by IR, (b) why can *in situ* atmospheric (stratospheric) O₃ be measured by UV.**

   a) Atmospheric CO_2 can be monitored by IR *in situ*, because CO_2 molecules are able to absorb IR radiation.

   b) The good ozone (O_3) in stratosphere (ozone layer) can be measured by UV-based *in situ* device because O_3 is known to strongly absorb UV radiation. (This is also the reason why ozone layer can protect humans from UV exposure.)

26) **Explain (a) why won't O₂ and N₂ affect the monitoring of CO in auto emission using IR. (b) Why is a long cell needed to monitor trace organic compounds using IR?**

   a) Both O_2 and N_2 are monoatomic molecules. They do not absorb infrared radiation. These molecules are symmetrical, no matter how much the covalent bonds are stretched there is no change in dipole moment. As a result, they will not interfere with CO measurement by IR.

   b) A sample cell with a long light path is needed for low concentration of organic compounds in the air. Based on Beer's law ($A = \varepsilon\, l\, C$), one way to increase absorbance A is to increase the light path (l) if the concentration (C) is low.

27) **Why is water a good solvent in Raman spectroscopy?**

   Water cannot be used in IR due to its intense absorption of IR, but water can be used as a solvent in Raman spectroscopy. This is because water is a weak Raman scatterer meaning that samples can be analyzed in their aqueous solutions.

28) **Under what circumstances will molecular vibrations be Raman active but not IR active and vice versa?**

   Both Raman scattering and IR absorption are the results of changes in vibration modes of molecules. A compound is IR active when the vibration modes of the molecule change the dipole moment, and a compound is Raman active when the vibrations change the polarizability. A vibration mode can be IR active, Raman inactive, and vice versa, however not at the same time. For example, homonuclear diatomic molecules such as N_2, O_2, H_2, and Cl_2 are found to be Raman active, but are not IR active. Water is a weak Raman scatterer, but water has strong absorption of infrared light.

Problems

1) **Calculate (a) the energy (in eV) of a 5.3-Å X-ray photon and (b) the energy (kJ/mol) of a 530-nm photon in a visible radiation. Given the following constants and unit conversion factors: 1 Å = 10^{-10} m; 1 nm = 10^{-9} m; 1 nm = 10 Å, 1 J = 6.26×10^{18} eV, Planck constant (h) = 6.63×10^{-34} J × s, speed of light (c) = 3×10^8 m/s.**

   a) Energy (in eV) of a 5.3-Å X-ray photon:

   $$E = h \times \frac{c}{\lambda} = 6.63 \times 10^{-34}\, J \times s \times \frac{3\,10^8 \text{ m/s}}{5.3\,\text{Å} \times \dfrac{10^{-10}\text{ m}}{1\,\text{Å}}} \times \frac{6.26\,10^{18} \text{ eV}}{1\,\text{J}} = 2349 \text{ eV}$$

   b) Energy (kJ/mol) of a 530-nm photon of visible radiation per photon:

   $$E = h \times \frac{c}{\lambda} = 6.63 \times 10^{-34}\, J \times s \times \frac{3\,10^8 \text{ m/s}}{530\,\text{nm} \times \dfrac{10^{-9}\text{ m}}{1\,\text{nm}}} \times \frac{10^{-3} \text{ kJ}}{1\,\text{J}} = 3.75 \times 10^{-22}\,\text{kJ}$$

Since there are 6.02×10^{23} photons per mole, the total energy in the unit of kJ/mole is:

$$3.75 \times 10^{-22} \text{ kJ per photon} \times \frac{6.02 \times 10^{23} \text{ photons}}{1 \text{ mole}} = 225.8 \text{ kJ/mole}$$

2) **Perform the following calculation: (a) $A = 0.012$, % $T = ?$ (b) $T = 95\%$, $A = ?$; (c) wavelength = 350 nm, wavenumber = ? cm^{-1}; (d) wavenumber = 4000 cm^{-1}, frequency = ? hertz.**

a) $A = -\log_{10} T$. Hence $0.012 = -\log_{10} T$, we can get: $T = 0.973$ (97.3%).

b) $T = 95\% = 0.95$, $A = -\log_{10} T = -\log_{10} 0.95 = 0.022$.

c) $\bar{v} \left[cm^{-1} \right] = \dfrac{10,000}{\lambda \left[\mu m \right]} = \dfrac{10,000}{350 \text{ nm} \times \dfrac{1 \ \mu m}{1,000 \text{ nm}}} = 28,571 \text{ cm}^{-1}$

d) $\lambda \left[\mu m \right] = \dfrac{10,000}{\bar{v} \left[cm^{-1} \right]} = \dfrac{10,000}{4000 \text{ cm}^{-1}} = 2.5 \ \mu m = 2.5 \times 10^{-6}$ m. The frequency (v) can be calculated

as: $v = \dfrac{c}{\lambda} = \dfrac{3 \times 10^{8} \text{ m/s}}{2.5 \times 10^{-6} \text{ m}} = 1.2 \times 10^{14} \dfrac{1}{s} = 1.2 \times 10^{14}$ hertz

3) **The absorbance of a 2-cm sample cell of a 10 ppm solution is 0.43, what would be the absorbance of a 2.0-cm cell of 15 ppm solution of the same chemical under the same condition?**

We use Beer's law: $A = \varepsilon l C$. Given the spectrometric condition for a specific reaction, the molar absorptivity ε is a chemical-specific constant. Therefore, we have: $\dfrac{A_1}{l_1 C_1} = \dfrac{A_2}{l_2 C_2} = \varepsilon$

Rearrange the above equation, we have: $A_2 = A_1 \dfrac{l_2 C_2}{l_1 C_1} = 0.43 \times \dfrac{1 \text{ cm} \times 15 \text{ ppm}}{2 \text{ cm} \times 10 \text{ ppm}} = 0.3225$

4) **At 373 nm, a 10 mg/L solution containing Cr(VI) has a transmittance of 17.8% when measured in 2.0 cm cell. What is the transmittance when measured in a 1.0-cm cell? Atomic weight of Cr = 52.**

- Convert mg/L into M: $(10 \text{ mg/L}) \times (1 \text{ g/1000 mg}) \times (1 \text{ mol/52 g}) = 0.000192$ M
- Convert T into A: $A = -\lg T = -\lg 0.178 = 0.75$
- Applying Beer's law: $A_1/A_2 = \varepsilon C_1 l_1 / \varepsilon C_1 l_1 = l_1/l_2$ for the same ε (same compound) but different concentrations (C)
- $0.75/A_2 = 2 \text{ cm}/1 \text{ cm}$, $A_2 = 0.375$
- Convert A into T: $T = 10^{-0.375} = 0.42$ (42%)

5) **A 2-cm cell of 0.003% (w/v, i.e., 0.003 g/100 mL) solution of $C_{12}H_{17}NO_3$ (MW = 223.3) has a λ_{max} of 280 nm with an absorbance of 0.92. What is the absorbance of the same chemical when the concentration is 0.005% using a 1-cm cell under the same wavelength? Report ε in 1/(M-cm).**

From Beer's law, we have: $\varepsilon = A/(l \times C)$. The molar concentration of $C_{12}H_{17}NO_3$ is:

$$C = \frac{0.003 \text{ g}}{100 \text{ mL}} \times \frac{1 \text{ mol}}{223.3 \text{ g}} \times \frac{1000 \text{ mL}}{1 \text{ L}} = 1.343 \times 10^{-4} \text{ mol/L}$$

The molar absorptivity can be calculated:

$$\varepsilon = \frac{A}{l \times C} = \frac{0.92}{2 \text{ cm} \times 1.343 \times 10^{-4} \text{ mol/L}} = 3424 \frac{1}{M \times cm}$$

$$\frac{A_1}{l_1 C_1} = \frac{A_2}{l_2 C_2} = \varepsilon$$

Rearrange the above equation, we have: $A_2 = A_1 \dfrac{l_2 C_2}{l_1 C_1} = 0.92 \times \dfrac{1 \text{ cm} \times 0.005\%}{2 \text{ cm} \times 0.003\%} = 0.767$

6) **The molar absorptivities (ε) of two compounds A and B were measured with pure samples at two wavelengths. From these data in the following table, determine the concentration (in M) of A and B in the mixture. The absorbance of a mixture of A and B in a 1.0-cm cell was determined to be 0.886 at 277 nm and 0.552 at 437 nm.**

	ε at 277 nm	ε at 437 nm
Chemical A	14,780 (M-cm)$^{-1}$	5,112 (M-cm)$^{-1}$
Chemical B	2,377 (M-cm)$^{-1}$	10,996 (M-cm)$^{-1}$

Applying Beer's law at two wavelengths, we can obtain two equations:

At 227 nm: $0.886 = 14780 \, M^{-1}cm^{-1} \times C_A \times 1 \, cm + 2377 \, M^{-1}cm^{-1} \times C_B \times 1 \, cm$

At 437 nm: $0.552 = 5112 \, M^{-1}cm^{-1} \times C_A \times 1 \, cm + 10996 \, M^{-1}cm^{-1} \times C_B \times 1 \, cm$

where C_A and C_B are the molar concentrations of chemical A and chemical B, respectively. By solving the above two simultaneous equations, we can obtain:

$C_A = 5.61 \times 10^{-5} M$

$C_B = 2.41 \times 10^{-5} M$

7) **Acetone has a molar absorptivity of 2.75×10^3 L $\times$ cm^{-1} $\times$ mol^{-1} at 366 nm, in ethanol. What range of concentrations in mg/L can be measured, so the absorbance remains within the linear range of 0.08–0.3, using a 1.0 cm cell (acetone has a molecular weight of 58)?**

- For $A = 0.08$: $C = A/\varepsilon l = 0.08/[(2.75 \times 10^3 \, L \times cm^{-1} \times mol^{-1}) \times 1 \, cm] = 2.9 \times 10^{-5}$ mol/L $= (2.9 \times 10^{-5}$ mol/L$) \times (58,000$ mg/mol$) = 1.68$ mg/L
- For $A = 0.3$: $C = A/\varepsilon l = 0.3/[(2.75 \times 10^3 \, L \times cm^{-1} \times mol^{-1}) \times 1 \, cm] = 1.1 \times 10^{-4}$ mol/L $= (1.1 \times 10^{-4}$ mol/L$) \times (58,000$ mg/mol$) = 6.38$ mg/L

8) **A solution of Co^{3+} shows a λ_{max} of 600 nm. (a) Calculate its splitting energy due to d-d electronic transition. (b) What are the absorbed color and the observed color?**

a) $\Delta E = \dfrac{hc}{\lambda} = \dfrac{\left(6.6210^{-34} \, JS\right) \times \left(3.0 \times 10^8 \, \dfrac{m}{s}\right)}{600 \, nm \times \dfrac{1 \, m}{10^9 nm}} = 3.31 \times 10^{-19}$ J

b) Using Table 9.1, we can tell that the absorbed color is orange, and the observed color is blue.

9) **Through calculation, demonstrate that the wavenumber corresponding to the stretching of C–H single bond is within the IR range (given $k = 5 \times 10^5$ g/s^2).**

$m_1(\text{for one C atom}) = \dfrac{12 \, g}{1 \, mol} \times \dfrac{1 \, mol}{6.02 \times 10^{23}} = 1.99 \times 10^{-23} g$

$m_1(\text{for one H atom}) = \dfrac{1 \, g}{1 \, mol} \times \dfrac{1 \, mol}{6.02 \times 10^{23}} = 0.167 \times 10^{-23} g$

For the bond between C–H, the reduced mass:

$\mu = \dfrac{m_1 m_2}{m_1 + m_2} = \dfrac{1.99 \times 10^{-23} \times 0.167 \times 10^{-23}}{1.99 \times 10^{-23} + 0.167 \times 10^{-23}} = 1.54 \times 10^{-24} g$

$\bar{v} = \dfrac{1}{2\pi c}\sqrt{\dfrac{k}{\mu}} = \bar{v} = \dfrac{1}{2 \times 3.14 \times 3.0 \times 10^8 m/s}\sqrt{\dfrac{5 \times 10^5 g/s^2}{1.54 \times 10^{-24} g}} = 302,400 \, m^{-1} = 3,024 \, cm^{-1}$

This wavenumber can be seen in Fig. 9.21 for C–H stretching. The range of variations in the wavenumber in the figure is due to the molecular environment of C–H group within the molecule.

10) **Nitrile (-C≡N) has an absorption band at 4.44 μm. What is the corresponding wave-number in cm^{-1}?**

Applying Eq. 9.7, we have $\bar{v} = 10000 / 4.44 = 2{,}250$ cm^{-1}.

11) **How many vibrational modes are possible for (a) NH_3, (b) H_2O, (c) CO_2, (d) CCl_4?**

Only CO_2 is a linear molecule, use $3N{-}5$: $3N{-}5 = 3 \times 3 - 5 = 4$

For all three other nonlinear molecules, use $3N{-}6$: $NH_3 = 3 \times 4 - 6 = 6$; $H_2O = 3 \times 3 - 6 = 3$; $CCl_4 = 3 \times 5 - 6 = 9$.

Chapter 10

Questions

1) **Explain: (a) The difference in the electronic configuration among various species of calcium, i.e., Ca in $CaCl_2$, Ca^0, Ca^{0*}, Ca^+, and Ca^{+*}. (b) Of these species, which one(s) are the desired for AAS measurement of Ca and which one(s) are unwanted species that may cause interference for AES measurement of Ca?**

   a) Ca in $CaCl_2$ is the Ca initially present in the salt form in the aqueous sample. After the removal of water (desolvation), a gaseous $CaCl_2$ is produced, which is further dissociated into gaseous atom (Ca^0). At an elevated temperature, Ca can have other electronic configurations such as excited vapor Ca atom (Ca^{0*}), ionic Ca (Ca^+), and ionic Ca with an excited electron (Ca^{+*}).

   Ca in $CaCl_2$: $1s^2\,2s^2\,2p^6\,3s^2\,3p^6$ (it is Ca^{2+} in solution, two valance electrons are lost)

   Ca^0: $1s^2\,2s^2\,2p^6\,3s^2\,3p^6\,4s^2$

   Ca^{0*}: $1s^2\,2s^2\,2p^6\,3s^2\,3p^6\,4s^2$ (same as Ca^0, but one electron is in its excited state)

   Ca^+: $1s^2\,2s^2\,2p^6\,3s^2\,3p^6\,4s^1$ (one valence electron remains, and second one is lost)

   Ca^{+*}: $1s^2\,2s^2\,2p^6\,3s^2\,3p^6\,4s^1$ (same as Ca^+, but the valence electron is in its excited state)

   b) Of these species, the gaseous Ca atom (Ca^0) is the only desired species for atomic absorption spectrometry (AAS), because the absorption is based on Ca^0. All other species are unwanted because they more or less cause interference. For atomic emission spectrometry (AES), the emission is based on the excited vapor Ca atom (Ca^{0*}) which is produced directly from Ca^0. When Ca^{0*} is emitted, it returns to its ground level as Ca^0. Any species beyond the formation of (Ca^{0*}) are unwanted and will cause interference during the AES measurement.

2) **Explain: (a) What fractions of atoms in the flame are typically in the excited states? (b) Why flame absorption-based spectrometers are normally more sensitive than flame emission spectrometers?**

   a) The ratio of excited vs ground state atoms in the flame is very low. As calculated by the Boltzmann equation (Example 10.1), this ratio for sodium is 2.50×10^{-4} at 2600 K and 2.59×10^{-4} at 2610 K. These ratios imply that the majority of the Na atoms (as well as for other elements) remain in their ground state.

   b) Since flame atomic absorption spectroscopy (FAAS) is based on much larger population of ground-state atoms, they are expected to be more sensitive than flame emission spectrometers.

Fundamentals of Environmental Sampling and Analysis, Second Edition. Chunlong Zhang.
© 2024 John Wiley & Sons, Inc. Published 2024 by John Wiley & Sons, Inc.
Companion Website: www.wiley.com/go/EnvironmentalSamplingandAnalysis2e

3) **Explain the difference in light source used in atomic absorption vs the light source used in UV-VIS spectrometer.**

   Light sources used is FAAS is a hollow cathode lamp (HCL) of the element being measured. In UV-visible spectrophotometry, a low-pressure hydrogen or deuterium lamp is used for UV and tungsten filament incandescent lamp for visible measurements.

4) **For the following atomic absorption spectrometers – FAAS, GFAAS, CVAAS: (a) sketch the schematic diagram; (b) describe the principles (functions) of major instrumental components.**

   Figures 10.3 and 10.4 can be referred to as FAAS and GFAAS. Figure 10.5 is a schematic diagram of CVAAS. Students' answers will vary for the description of components of these instruments. Below is a brief description of atomic absorption spectroscopy (AAS).

   All atomic absorption (AA) spectrometers (FAAS, GFAAS, CVAAS) use the absorption of light to measure the concentration of gas-phase atoms. Since samples are usually liquids or solids, the analyte atoms or ions must be vaporized first in a flame, graphite furnace, or cold vapor AAS. The vapor atoms then absorb ultraviolet or visible light and make transitions to higher electronic energy levels. The analyte concentration is determined from the amount of absorption based on Beer's law.

   Light source: The light source is usually a hollow-cathode lamp of the element that is being measured. Because of the need for an element-specific narrow-band light source, all atomic absorption spectrometers are disadvantageous (compared to ICP) in that only one element is measurable at a time.

   Atomizer: AA spectroscopy requires that the analyte atoms be in the gas phase. Ions or atoms in a sample must undergo desolvation and vaporization in a high-temperature source such as a flame or graphite furnace (FAAS and GFAAS), or vaporized into a vapor form of mercury under room temperature condition (CVAAS). FAAS and CVAAS can only analyze solutions, while graphite furnace AAS can accept solutions, slurries, or solid samples. Flame AAS uses a slot-type burner to increase the path length, thereby increasing the total absorbance. Sample solutions are usually aspirated with the gas flow into a nebulizing/mixing chamber to form small droplets before entering the flame. The graphite furnace is a much more efficient atomizer than a flame and it can directly accept very small absolute quantities of samples. It also provides a reducing environment for easily oxidized elements. Samples are placed directly in the graphite furnace and the furnace is electrically heated in several steps to dry the sample, ash organic matter, and vaporize the analyte atoms.

   Light separation and detection: AA spectrometers use monochromators and detectors for UV and visible light measurement. The main purpose of the monochromator is to isolate the absorption line from background light due to interferences. Simple dedicated AA instruments often replace the monochromator with a bandpass interference filter. For the subsequent measurement of UV and visible light measurement, photomultiplier tubes are the most common detectors for AA spectroscopy by converting photon signals into electrical signals.

5) **For the following atomic emission spectrometers – flame atomic emission, ICP-OES, and XRF: (a) sketch the schematic diagram; (b) describe the principles (functions) of major instrumental components.**

   Figures 10.7 and 10.8 can be referred to as ICP-OES and XRF, respectively. Students' answers will vary for the description of components of these instruments. Please refer to the book for the description of instrumental principles and components.

6) **Explain what circumstance makes an axial view or radial view more appropriate in ICP-OES.**

In the radial view, only a small angle section of the light is used, which results in the highest upper linear ranges. In the axial view, the plasma is viewed (by the spectrometer) along the length of the plasma. The axial view reduces the continuum background from the ICP itself but maximizes the sample path; it provides better detection limits – by as much as a factor of 10, than those obtained by radial ICP–OES.

7) **Briefly answer the following questions: (a) mechanism of hollow cathode lamp; (b) atomization procedure in graphite furnace; (c) principles of cold vapor mercury analyzer; (d) principles for the formation of plasma; (e) how X-ray fluorescence is produced in XRF.**

a) Hollow cathode lamp is used as a light source in AAS. Its cathode is made of element of interest with a low pressure of an inert gas. A low electrical current is imposed in such a way that the metal is excited and emits a few spectral lines characteristic of that element.

b) In GFAA, instead of using an aspiration device, either liquid or solid form of samples is deposited directly to a graphite boat. Elements in the samples are atomized through the heating of the graphite furnace which is temperature programmed.

c) In a cold vapor mercury analyzer, mercury of various chemical forms is chemically reduced to free atomic state (Hg^0) by reacting the sample with a strong reducing agent. The volatile free mercury is driven from the reaction flask by bubbling air through the solution. Mercury atoms are carried in the gas stream to an absorption cell, which is placed in the light path of a spectrometer. As the mercury atoms pass into a sampling cell, measured absorbance rises indicating the increased concentration of mercury atoms in the light path.

d) Plasma is formed in an ICP plasma torch. A typical plasma torch consists of concentric quartz tubes. The inner tube contains the sample aerosol and argon (Ar) support gas and the outer tube contains flowing gas to keep the tubes cool. A radio frequency (RF) generator (typically 1–5 kW at 27 MHz) produces an oscillating current in an induction coil that wraps around the tubes. The induction coil creates an oscillating magnetic field, which in turn sets up an oscillating current in the ions and electrons of the support gas (Ar). The energy transferred to a stream of argon through an induction coil produces temperatures up to 10,000 K plasma. This high-temperature plasma atomizes the sample and promotes atomic and ionic transitions, which are observable at UV and visible wavelengths.

e) As shown in Fig. 10.8, the energy dispersive XRF consists of a polychromatic source (either an X-ray tube or a radioactive material), a sample holder, a semiconductor detector, and the various electronic components required for energy discrimination. The emitted photons released from the sample are observed from the sample at 90^0 angle to the incident X-ray beam so that the incident light will not interfere with the detector. In the detector, each photon strikes a silicon wafer that has been treated with lithium and generates an electrical pulse that is proportional to the energy of the photon. The concentration of the element is determined by counting the number of pulses.

8) **Describe (a) nebulizer and atomizer in the FAAS; (b) is there a nebulizer in GFAAS?**

a) In an FAA system (Fig. 10.4A), the *nebulizer* is to suck up liquid sample at a controlled rate, create a fine stream of aerosols, and mix the aerosols with fuel oxidant thoroughly for introduction into the flame. In an *atomizer*, the flame destroys any analyte ions and breakdown complexes, and creates atoms (the elemental form) of the element of interest into vapor atoms such as Fe^0, Cu^0, Zn^0.

b) In a flameless graphite furnace system (Fig. 10.4B), there is no nebulizer device. Instead of using an aspiration device, both liquid and solid samples can be loaded directly to a graphite boat using a syringe inserted through a cavity. The graphite furnace can hold an atomized sample in the optical path for several seconds, compared to only a fraction of a second in the flame system. This results in a significantly higher sensitivity of the GFAAS as compared to FAAS.

9) **Briefly describe: (a) advantages of GFAAS over FAAS; (b) why CVAAS rather than FAAS should be particularly useful for Hg analysis; (c) advantages of ICP over flame AAS; (d) principal advantages and drawback of XRF and why XRF is unique among all atomic spectroscopic techniques.**

a) In FAAS, only about 10% of the sample will reach the burner and be atomized, whereas the entire sample is atomized in the GFAAS. Furthermore, GFAAS can hold the atomized sample in the optical path for several seconds, whereas with the flame system, the sample is held for only a fraction of a second. Hence in theory, GFAAS should be approximately 1,000 times more sensitive than the regular FAAS.

b) Mercury (Hg) is a volatile metal, so it does not need to be heated in a flame or furnace during spectroscopic analysis. It can be analyzed by a "cold-vapor" atomic absorption (CVAAS) technique.

c) The most important advantage of ICP over FAAS is its high sample throughput. ICP can analyze more elements and these elements are almost simultaneously measured after a sample is introduced. FAAS can only analyze one element at a time. A hollow cathode lamp needs to be changed when another element is measured.

d) XRF is unique in that it is a nondestructive analysis compared to FAAS techniques. It can analyze most heavy metals and can be easily operated. However, the detection limit is poor.

10) **In comparing the flame atomic absorption spectroscopy (FAAS) and the electrothermal (graphite furnace) atomic absorption spectroscopy (GFAAS): (a) explain why atomization efficiency is very low for FAAS; (b) explain why GFAAS is much more sensitive than FAAS; (c) explain why GFAAS analysis usually is slower (2–3 min per sample) than FAAS (few seconds); (d) what is the major problem with GFAAS.**

a) and b) In FAAS, only about 10% of the sample will reach the burner and be atomized, whereas the entire sample is atomized in the GFAAS. However, GFAAS can hold the atomized sample in the optical path for several seconds, whereas with the flame system, the sample is held for only a fraction of a second. Hence in theory, GFAAS should be approximately 1,000 times more sensitive than the regular FAAS.

c) This is because GFAAS holds sample and the furnace is programmed to undergo a sequential temperature change to remove solvent and matrix components prior to atomization, whereas for FAAS, a liquid sample is continuously introduced (pumped) into a nebulizer and flame torch and the absorbance is measured instantaneously.

d) The main disadvantages of GFAAS are its slow analysis rate and its limited dynamic range. Its linear range of response is very dynamically limited. In general, its absorbance is directly proportional to analyte element concentration for solutions ranging from one part per billion (1 ppb) to 100 ppb. This represents only two orders of magnitude. Above this range the absorbance begins to taper off and is no longer linearly related to concentrations.

11) **Briefly define what are (a) spectral interference, (b) chemical interference, and (c) physical interference.**

a) Spectral interference: The spectral interference occurs in both atomic emission and atomic absorption spectroscopy. In atomic emission, the spectral interference occurs when either another emission line or a molecular emission band is close to the emitted line of the test element and is not resolved from it by the monochromator. Such molecular emission could come from the oxides of other elements in the sample. In atomic absorption techniques, solid particles, unvaporized solvent droplets, or molecular species in the flame can cause a positive spectral interference.

b) Chemical interference: The chemical interference is a result of the formation of undesired chemical species during the atomization process, such as ions and refractory compounds. Their effects are more common than spectral ones, but can be frequently minimized by selecting a proper operating condition. Chemical interferences are more common in low-temperature systems such as FAAS and GFAAS than in high-temperature ICP systems.

c) Physical interference: The physical interference is caused by the variation of instrumental parameters that affect the rate of sample uptake in the burner and atomization efficiency. This includes variations in the gas flow rates, variation in sample viscosity due to temperature or solvent, high solids content, and changes in flame temperature. Physical interferences can be corrected by the use of internal standards.

12) **The Standard Method 3500-Li B determines lithium (Li) at a wavelength of 670.8 nm using a flame emission photometric method. In the interference section, it states: "A molecular band of strontium hydroxide with an absorption maximum at 671.0 nm interferes in the flame photometric determination of lithium. Ionization of lithium can be significant in both the air-acetylene and nitrous oxide-acetylene flames and can be suppressed by adding potassium." (a) Describe the spectral interference of this method. (b) Describe why lithium ionization is a problem and how the addition of potassium can mitigate such a problem.**

a) The spectral interference is due to the overlap of two very close emission lines, 670.8 nm for Li and 671.0 nm for strontium hydroxide.

b) Lithium is an easily ionizable element. The ionization will be detrimental to an emission-based spectroscopy. Potassium has a lower first ionization energy than lithium. (The outermost electrons of a potassium atom are farther from its nucleus than the outermost electrons of a lithium atom are from their nucleus. So, the outermost electrons of a lithium atom are held more tightly to its nucleus. As a result, removing an electron from a potassium atom takes less energy than removing one from a lithium atom.) Therefore, the addition of potassium will suppress the ionization of lithium.

13) **Sulfate (SO_4^{2-}) and phosphate (PO_4^{3-}) ions hinder the atomization of Ca^{2+} by the formation of nonvolatile salts. Explain why EDTA (ethylenediaminetetraacetic acid) can be added to protect Ca measurement from SO_4^{2-} and PO_4^{3-} interference.**

The presence of phosphates and sulfates will interfere with calcium measurements due to the formation of refractory calcium. These refractory compounds cannot be atomized in flames or plasmas. To prevent the formation of refractory compounds, one can add a releasing agent, such as EDTA (ethylenediaminetetraacetic acid), because EDTA is a strong chelating agent.

14) **Explain why cesium (Cs) salt should be used to minimize the ionization interference when analyzing K and Na using FAAS.**

Both K and Na are easily ionizable elements; the ionization will decrease the sensitivity for both absorption- and emission-based measurements. For example, K and Na in their *ionized* forms cannot emit their characteristic *atomic* emission lines. Cesium is more easily ionizable than K and Na, so it can serve as the ionization suppressant for K and Na. This is because ionized K or Na will capture electrons released from cesium and revert ionized K and Na back to their natural atomic forms. Ionization interferences are typical for measuring easily ionizable Group 1 and 2 elements.

15) **Explain how Zeeman background correction is used to minimize spectral interference.**

Zeeman background correction has been widely used in GFAAS because the background measurement is made at the analytical wavelength rather than other wavelengths. This method uses a magnetic field around the atomizer. The field splits the energy levels of the absorbing atoms and allows discrimination of atomic absorption from other sources of absorption.

16) **Explain why NH_4NO_3 is added to seawater when Pb and Ca are analyzed by GFAAS. (*Hint*: $NH_4NO_3 + NaCl \rightarrow NH_4Cl + NaNO_3$.)**

NH_4NO_3 is to serve as a chemical modifier in GFAAS. Its addition to seawater will significantly increase the volatility of seawater matrix component (mainly NaCl) according to the reaction: $NH_4NO_3 + NaCl \rightarrow NH_4Cl + NaNO_3$.

17) **Explain why (a) HCl cannot be used as a sample matrix when Cu, Pb, Zn, Ca are analyzed using GFAAS; (b) why moisture needed to be removed in cold vapor gas cell.**

a) HCl should be avoided as a sample matrix if Cu, Pb, Zn, Ca are analyzed, because spectral interference may occur due to the presence of chloride. Chloride interference may be mitigated by adding ammonium salts, which cause the volatilization of NH_4Cl and hence the removal of chloride.

b) Under the room temperature condition of the cold vapor atomic absorption spectrometry, moisture will cause condensation in the absorption cell and interfere with the measurement of mercury.

18) **What are the major factors (criteria) in selecting which atomic spectroscopic instruments for metal analysis?**

The major factors in selecting which atomic spectroscopic instruments for metal analysis are detection limits, analytical working range, sample throughput, cost, interferences, ease of use, and availability of proven methodology.

19) **Generally speaking, which of the following has the lowest (best) detection limits, the widest linear range, the fastest sample throughput, or the lowest purchasing price, respectively: (a) FAAS; (b) GFAAS; (c) ICP-OES, (d) ICP-MS.**

Generally, ICP-MS has the lowest detection limit. ICP-OES has the widest working range. ICP-OES and ICP-MS are the fastest technique among these because of their capacity for sequential/simultaneous measurement of multiple elements. FAAS is the low-cost alternative if only a few fixed element(s) are routinely determined.

20) **Explain: (a) why XRF is particularly useful as a tool in screening tests of contaminated sites; (b) what problem occurs if flame atomization is used for mercury analysis.**

a) XRF is useful as a screening tool because XRF is often made as a handheld portable device, which adds convenience to *in situ* uses, such as soil metal contamination, Brownfield site investigation, and metal testing in industrial hygiene testing. However, because of its low sensitivity, it can only be used as a "screening" tool rather than accurate determination of metals.

b) Elemental mercury is volatile at room temperature. With flame ionization, mercury vapor will not be able to reach the absorption cell.

21) **ICP-OES has become increasingly popular in environmental labs for metal analysis, and it seems that the flame and flameless AA techniques are diminishing. (a) Explain why; (b) list major advantages and disadvantages of ICP-OES for elemental (metal) analysis.**

a) ICP-OES has become the dominant instrument for routine metal analysis primarily because of its ability in performing multielement analysis in a sequential/simultaneous mode. Other metal analysis methods (FAAS, GFAAS) are commonly seen as a single element method, because a hollow cathode lamp (HCL) needs to be changed whenever the element of interest is changed.

b) Besides its excellent sample throughput, ICP-OES has several other advantages, including: (1) It has lower interelement interferences because of the higher temperatures. (2) ICP-OES gives a much wider linear dynamic range of 4–6 orders of magnitude than its counterparts. However, ICP-OES has some disadvantages as well. ICP spectra are complicated and made up of hundreds of lines. This requires great resolution and more sophisticated equipment (more expensive than FAAS and GFAAS). It also has higher operating costs due to its huge argon gas consumption. Skilled operators are necessary to develop the rather complicated methods. ICP can also be less precise than AAS in some applications. A vacuum UV spectrometer is required to "see" B, P, N, S, and C and there is limited use for the alkali metals since their lines approach near-IR wavelengths.

22) **Explain why is ICP-OES particularly troublesome for the analysis of several environmentally important metals and metalloids including As, Se, Hg, and Cr.**

ICP-OES can analyze most metals, but one downside is that it is not very sensitive to As, Se, Cr, and Hg because of the issues associated with the variable sensitivity dependent on the oxidation state, the volatile nature, or the high carryover potentials.

23) **(a) Sketch and briefly describe the hydride generation atomic absorption (HGAAS) technique for the analysis of arsenic and selenium. (b) What are the operational parameters that affect the sensitivities?**

a) Figure 10.6 can be referenced for a schematic of HGAAS for the analysis of arsenic and selenium. It typically consists of: (1) a hollow cathode lamp (to provide the analytical light line for the element of interest and provide a constant yet intense beam of that analytical line); (2) a hydride generation system (to aspirate liquid sample at a controlled rate, mix liquid sample with sodium borohydride and HCl, create a volatile hydride of the analyte metalloid from that reaction, and flow that gaseous hydride into the optical cell); (3) optical cell and flame (to decompose the hydride form of the metalloid from the hydride generation module, thereby creating atoms of the element of interest); (4) a monochromator (to isolate analytical lines' photons passing through the optical cell, remove scattered light of other wavelengths from the optical cell); and (5) a photomultiplier tube, PMT (as the detector, the PMT determines the intensity of photons of the analytical line exiting the monochromator).

b) Results depend heavily on a variety of parameters, including the valence state of the analyte, reaction time, gas pressures, acid concentration, and cell temperature. Therefore, the success of the HGAAS technique will vary with the care taken by the operator in attending to the required detail. The formation of the analyte hydrides is also suppressed by a number of common matrix components, leaving the technique subject to certain chemical interferences.

24) **In the US EPA's SW-846 methods: (a) describe two multielement atomic spectrometric methods; (b) describe two most common single element atomic AA; (c) what else atomization techniques are used for several other elements such as Hg, As, and Se?**

   a) Two multielement atomic spectrometric methods in SW-846 are: Method 6010C – Inductively Coupled Plasma-Atomic Emission Spectrometry; Method 6200 – Field Portable X-Ray Fluorescence Spectrometry for the Determination of Elemental Concentrations in Soil and Sediment.

   b) Two common single-element atomic AA are FAAS and GFAAS, including Method 7000B – Flame Atomic Absorption Spectrophotometry; Method 7010 – Graphite Furnace Atomic Absorption Spectrophotometry.

   c) Besides the above-mentioned FAAS, GFAAS, ICP, XRF methods, Hg, As, and Se have different atomization techniques. These include: (1) methods for Hg: Method 7470A – Mercury in Liquid Waste (Manual Cold-Vapor Technique); Method 7471B – Mercury in Solid or Semisolid Waste (Manual Cold-Vapor Technique); (2) methods for As: Method 7061A – Arsenic (Atomic Absorption, Gaseous Hydride); Method 7062 – Antimony and Arsenic (Atomic Absorption, Borohydride Reduction); and (3) methods for Se: Method 7741A – Selenium (Atomic Absorption, Gaseous Hydride), Method 7742 – Selenium (Atomic Absorption, Borohydride Reduction).

25) **What are the two most commonly used hyphenated instrumental methods for metal species analysis?**
   HPLC-ICP-MS and CE-ICP-MS

26) **In flameless graphite furnace method, which quantitative method is more appropriate: (a) internal standard method; (b) standard addition methods? Explain.**
   The *standard addition method* can be used for the correction of certain interferences using GFAAS. The *internal standard method* typically requires multielement instrument techniques such as ICP-OES, because the internal standard and the element of the interest (analyte) should be measured at the same time. Additional explanations are provided below.

   Standard addition method: The standard addition method involves the addition of various amounts of analyte (as standard) to aliquots of the sample to be measured. A graph is plotted between the absorbance (*y*-axis) vs the mass of analyte added as standard (*x*-axis). Linear regression is used to define a mathematical relationship (line) between absorbance and mass. The presence of a nonzero absorbance at zero-added analyte is attributive to the analytes present in the unknown sample. The standard addition method can compensate for matrix effect (chemical interference). It does not account for interferences caused by the absorbance, emission, or scattering of light by nonanalyte (matrix) components (e.g., spectral interferences).

   Internal standard method: This method uses the ratio of the detector response of the analyte to response of a second element (the internal standard) that has been added to the sample being measured. The instrument should be capable of measuring these two elements simultaneously (such as ICP, but not the most common atomic spectrometric instruments). A suitable internal standard should have similar chemical and spectral properties to the analyte, and hence this technique will compensate for a change in atomizer condition or the presence of spectral interference. This method can also compensate for chemical interference if both the analyte and internal standard are equally affected by an interfering chemical.

27) **In Table 10.2, 10 samples were run as a batch along with 6 calibration standards. Out of a total of 26 runs, how many QA/QC samples were placed in this run sequence? What precautions regarding QA/QC should be made in running a batch of samples in the automatic sample log (sequence table)?**

In the sequence listed in Table 10.2, 10 QA/QC samples, 6 standards, and 10 samples are run. QA/QC samples account for $10/26 = 38\%$ of all the samples during the analysis. It is essential to intermittently place various types of QA/QC samples to ensure contamination is absent using various blanks (e.g., calibration blank, method blank), accuracy and precisions are acceptable through various standards and spikes (e.g., calibration verification standard, matrix spike, and matrix spike supplicate, etc.). Doing so, the analyst will have confidence in the data quality, and if things go wrong, it will make it easier to diagnose the problem and fix the issue.

28) **What is (are) method(s) of standard calibration suitable to minimize matrix effects?**

Among the three methods described in the textbook (external standard calibration, internal standard method, standard addition calibration), the standard addition method is suitable to minimize the matrix effects. In this method, a series of standard solutions containing various volumes (V_s) of the analyte is added to a constant volume (V_u) of sample containing the same analyte at an unknown concentration (C_u) to be determined. Each volumetric flask is diluted with solvent to the same final volume (V_f). Since the analytes in the standard experience the same matrix effects as the analyte in the sample, this method accounts for the sample matrix effects that may be the result of various factors (e.g., interference, high viscosity, or chemical reaction with the analyte).

Problems

1) **A groundwater sample is analyzed for its K by FAAS using the method of standard additions. Two 500 μL of this groundwater sample are added to 10.0-mL DI water. To one portion, 10.0 μL of 10 mM KCl solution was added. The net emission signals in arbitrary units are 20.2 and 75.1. What is the concentration of K in this groundwater in mg/L and mM?**

If we assume the concentration of K in groundwater is x (mM, or μmol/mL), then the concentration of K in 10.0-mL DI water spiked with 500 μL (0.5 mL) groundwater is:

$$C_1 = \frac{0.5 \text{ mL} \times x \dfrac{\mu\text{mol}}{\text{mL}}}{10 \text{ mL} + 0.5 \text{ mL}} = \frac{0.5x \ \mu\text{mol}}{10.5 \text{ mL}} = \frac{0.5x}{10.5} \frac{\mu\text{mol}}{\text{mL}}$$

Similarly, the concentration of K in 10.0-mL DI water spiked with 500 μL (0.5 mL) groundwater and 10 μL (0.01 mL) 10 mM (i.e., 10 μmol/mL) is:

$$C_2 = \frac{0.5 \text{ mL} \times x \dfrac{\mu\text{mol}}{\text{mL}} + 0.01 \text{ mL} \times 10 \dfrac{\mu\text{mol}}{\text{mL}}}{10 \text{ mL} + 0.5 \text{ mL} + 0.01 \text{ mL}} = \frac{(0.5x + 0.1) \ \mu\text{mol}}{10.51 \text{ mL}} = \frac{0.5x + 0.1}{10.51} \frac{\mu\text{mol}}{\text{mL}}$$

Since the emission signal is proportional to the concentration, we have:

$$\frac{C_1}{C_2} = \frac{0.5x/10.5}{(0.5x + 0.1)/10.51} = \frac{20.2}{75.1}$$

Solve for $x = 4.24 \times 10^{-2}$ μmol/mL $= 4.24 \times 10^{-2}$ mM

To convert this into mg/L, we use the atomic weight of 39 for K:

$$K\left(\frac{mg}{L}\right) = 4.24 \times 10^{-2} \frac{\mu mol}{mL} \times \frac{39\ \mu g}{1\ \mu mol} = 1.65 \frac{\mu g}{mL} = 1.65 \frac{mg}{L}$$

2) **A 5-point calibration curve was made for the determination of Pb in FAAS. The regression equation was: $y = 0.155\,x + 0.0016$, where y is the signal output as absorbance, and x is the Pb concentration in mg/L. (a) A contaminated groundwater sample was collected, diluted from 10 mL to 50 mL, and analyzed without digestion. The absorbance reading was 0.203 for the sample. Calculate the concentration of Pb in this groundwater sample. (b) A sediment sample suspected of Pb contamination was collected. After decanting the overlying water, a 1-g wet sediment sample was digested and diluted to 50 mL. FAAS measurement gave an absorbance reading of 0.350. A subsample of this wet sediment was taken to measure the moisture content with oven drying overnight. The moisture was 35%. Report the Pb concentration in sediment sample on a dry basis.**

a) Using the calibration curve, we have: $0.203 = 0.155\,x + 0.0016$. Solve for $x = 1.30$ mg/L. Since this sample was diluted 5 times (from 10 mL to 50 mL) prior to analysis, the original concentration of Pb in contaminated groundwater $= 1.30 \times 5 = 6.50$ mg/L

b) Using the calibration curve, we have: $0.350 = 0.155\,x + 0.0016$. Solve for $x = 2.25$ mg/L. Note that 2.25 mg/L is the Pb concentration in the digested solution. This needs to be converted into the initial concentration in sediment with the unit of mg/kg as follows:

$$\text{Pb in sediment}\left(\frac{mg}{kg}\right) = \frac{2.25 \frac{mg}{L} \times 50\ mL \times \frac{1\ L}{1000\ mL}}{1\ g \times \frac{1\ kg}{1000\ g}} = 112.5 \frac{mg}{kg}\ \text{(wet basis)}$$

Then we convert the wet basis to dry basis using Eq. 2.4 in Chapter 2:

$$Pb\left(\frac{mg}{kg}\ \text{on dry basis}\right) = \frac{\frac{mg}{kg}\ \text{on wet basis}}{1 - \%\ \text{moisture}} = \frac{112.5}{1 - 0.35} = 173.1 \frac{mg}{kg}\ \text{(dry basis)}$$

3) **The concentration of copper (Cu) in industrial wastewater was analyzed directly with ICP-OES following filtration to remove suspended solids. A subsample of 10 mL was diluted to a final volume of 25 mL with DI water and analyzed by ICP-OES, the instrumental signal was 2350. A second subsample of 10 mL was spiked with 10 μL of a standard solution containing 150 mg/L Cu and then diluted to 25 mL in a 25 mL volumetric flask. The ICP-OES signal was 7585. Report the Cu concentration in mg/L in the wastewater.**

We can apply Eq. 10.12 to solve for C_u using $V_S = 10\ \mu L = 0.010$ mL, $V_u = 10$ mL, and $C_S = 150$ mg/L:

$$C_u = \frac{y_1}{y_2 - y_1} \frac{V_S}{V_u} \times C_S = \frac{2350}{7585 - 2350} \frac{0.010}{10.0} \times 150 = 0.067 \frac{mg}{L}$$

4) **A standard solution containing 50 mg/L of lead (Pb) at various volumes (0–20 mL) was added to a constant volume (10 mL) of an unknown water sample. Following the analysis of Pb by GFAAS, the calibration curve was obtained through regression: $y = 0.0526\,V_s + 0.2743$, where y = instrument signal, V_S = volume of standard solution (mL). Determine the concentration of Pb in this water sample.**

Since the volume of standard solution (V_S) is given in the regression equation, Eq. 10.9b can be used, where slope (m') $= 0.0526$, intercept (b') $= 0.2743$.

$$C_u = \frac{b'}{m'} \frac{C_S}{V_u} = \frac{0.2743}{0.0526} \times \frac{50}{10} = 26.07 \frac{mg}{L}$$

Chapter 11

Questions

1) **Define the following forms of chromatography: GSC, GLC, LSC, LLC. Which one of these four chromatographic techniques is commonly referred to as GC? Which one is commonly referred to as LC?**

In GC, separation is based mainly on the partitioning between a gas mobile phase and a liquid stationary phase (*GLC, gas–liquid chromatography*). Sometimes separation is based on a gaseous or volatile compounds' adsorption ability on a solid stationary phase (*GSC, gas–solid chromatography*). GLC is typically correlated to the volatility (or boiling point) of the compound to be separated. A more volatile compound with a lower boiling point will be eluted first.

The partition-based gas–liquid chromatography (GLC) is commonly shortened to gas chromatography (GC). GLC is based on the partition of an analyte between a gaseous mobile phase and a liquid stationary phase immobilized on the surface of an inert solid. The gas–solid chromatography (GSC) has a stationary phase made of alumina (Al_2O_3) or porous polymers in the PLOT columns. The GSC is based on the adsorption of gaseous chemicals on the stationary phase. Owing to the semipermanent and the nonlinear sorption nature, the applications of GSC are limited to certain low-molecular-weight gaseous species such as components in air, H_2S, CS_2, CO, CO_2, and rare gases.

In LC, separation is often achieved on the basis of molecular polarity via partitioning between a liquid mobile phase and a liquid film adsorbed on a solid support material (*LLC, liquid–liquid chromatography*), and in some cases through adsorption on a solid stationary phase (*LSC, liquid–solid chromatography*).

Most of the liquid chromatographic systems are performed by LLC, which is based on the partition between liquid mobile phase and a liquid stationary film attached to a support surface (packing material). The liquid stationary film can be retained on surface through physical adsorption or by chemical bonding (the bonded phase). LSC has limited choices of stationary phases, using either silica or alumina, and is best suited to samples that are soluble in nonpolar solvents (i.e., low solubility in aqueous solvents). One particular application of LSC is its ability to differentiate isomeric mixtures.

Fundamentals of Environmental Sampling and Analysis, Second Edition. Chunlong Zhang.
© 2024 John Wiley & Sons, Inc. Published 2024 by John Wiley & Sons, Inc.
Companion Website: www.wiley.com/go/EnvironmentalSamplingandAnalysis2e

2) **Describe (a) the difference between adsorption-based chromatography and partition-based chromatography; and (b) the common stationary phases for GLC and LLC.**

    a) The GSC is based on the <u>*adsorption*</u> of gaseous chemicals on the stationary phase. Owing to the semipermanent and nonlinear sorption nature, the GSC is limited to certain low-molecular-weight gaseous species such as components in air, H_2S, CS_2, CO, CO_2, and rare gases. The GLC is based on the *partition* of analyte between a gaseous mobile phase and a liquid stationary phase immobilized on the surface of an inert solid.

    b) Common stationary phases for GLC in gas chromatography include polydimethyl siloxane, polyethylene glycol, and trifluoropropyldimethyl to name a few (Table 11.1, page 340). The LLC stationary phase in liquid chromatography is based on the partition between liquid mobile phase and a liquid stationary film attached to a support surface. In normal phase HPLC (NP-HPLC), the stationary phase is highly polar, such as cyano [$(CH_2)_3CN$], diol [$(CH_2)_2]CH_2(OH)CH_2OH$], amino [$(CH_2)_3NH_2$], and dimethylamino [$(CH_2)_3N(CH_3)_2$]. In reverse-phase HPLC (RP-HPLC), the stationary phase is a nonpolar liquid. The most common nonpolar bonded phases are hydrocarbons such as *n*-decyl ($C_8 = C_8H_{17}$) or *n*-octyldecyl ($C_{18} = C_{18}H_{37}$). C_8 is intermediate in hydrophobicity and C_{18} is very nonpolar.

3) **What differs between three capillary columns: WCOT, SCOT, and PLOT?**

WCOT is a wall-coated open-tubular capillary column. Its stationary phase is a thin liquid film that is coated on the wall of the capillary column. SCOT is a support-coated open-tubular capillary column. Its liquid stationary phase is coated onto microparticles, which are attached to the wall of the capillary. PLOT is a porous layer of open-tubular capillary column. Its solid-phase microparticles are attached to the capillary walls. WCOT and SCOT columns are used in partition-based gas–liquid chromatography (GLC) and PLOT columns are used in adsorption-based gas–solid chromatography (GSC).

4) **What differs between packed columns and capillary columns regarding their performance in separation? Why?**

A *packed column* is typically a glass or stainless-steel coil (1–5 m total length and 5 mm i.d.) that is filled with a stationary phase, or a packing coated with the stationary phase. A *capillary column* is a thin fused-silica (purified silicate glass SiO_2) tube (typically 10–100 m in length and 250 μ i.d.) that has the stationary phase inside. Capillary columns have much smaller diameters than packed columns. They have an open tubular structure without packing materials, but the stationary phases are coated by one of the three ways (WCOT, SCOT, PLOT).

5) **List the likely intermolecular force(s) for GC columns with (a) polysiloxane and (b) polyethylene glycol stationary phases.**

(a) dispersion, (b) dispersion, dipole–dipole, and H-bonding.

6) **What column (DB-1 or DB-WAX) one would choose for the GC analysis of (a) hydrocarbons, (b) ketones?**

DB-1 (nonpolar) columns for hydrocarbons; DB-WAX (polar) columns for ketones.

7) **Four alkanes are analyzed by GC with HP-1 as the column: *n*-butane, propane, hexane, and *n*-pentane. Predict the elution order and explain why. What intermolecular force will predominate?**

HP-1 column is a column with nonpolar 100% dimethylpolysiloxane. The four low molecular weight homologous series hydrocarbons are also nonpolar, with increasing molecular weights, boiling points, and hydrophobicities in an order of propane < *n*-butane < *n*-pentane < hexane. When a nonpolar HP-1 column is used to separate nonpolar hydrocarbons, the London force becomes the only intermolecular force (IMF). The London dispersion force is an attractive force due to the formation of temporary (short-lived) dipoles in a nonpolar molecule. When the electrons in two adjacent molecules are displaced in such a way that molecules gain some temporary dipoles, they attract each other through the weak London dispersion force. The elution order will be the same: propane < *n*-butane < *n*-pentane < hexane.

8) **Relate each term in van Deemter equation (i.e., A, B, and C) to: (a) random molecular diffusion, (b) mass transfer, (c) eddy diffusion.**

   A = Eddy diffusion, B = random molecular diffusion, C = mass transfer

9) **Describe (a) why is temperature important for a good separation in GC, but not mobile phase; (b) why is solvent mobile phase important for a good separation in HPLC but not temperature.**

   a) In GC, various compounds are separated in the column according to their relative volatilities. As column temperature increases in the oven, a compound with a lower boiling point will become volatile first and escape the column. The role of a mobile phase (mostly helium) in the GC is to carry the compound from the column to the detector, hence the name carrier gas.

   b) In HPLC, on the contrary, temperature has a little effect on separation. In fact, room temperature is commonly used. However, the composition of the mobile phase is extremely important because separation in HPLC is based entirely on the polarity of mobile phase solvents. Hence, from the separation standpoint, temperate to GC is as to mobile phase to HPLC.

10) **Describe the effect of the following variables on the resolution and analytical speed (length of retention time) in GC: (a) column diameter, (b) film thickness, and (c) column length.**

    a) A larger diameter column will decrease the analytical speed, but will increase the resolution.

    b) A thicker film of stationary phase will decrease analytical speed, but its effect on the resolution depends on boiling points of analytes in GC.

    c) A longer column will decrease the analytical speed, but it will increase resolution. In theory, doubling the column length will increase the resolution by 1.4 times.

11) **List three methods to improve the resolution in HPLC.**

    Selecting a mobile phase (solvents) with polarity adequate for the analytes of interest, changing flow regime from isocratic to gradient flow, and using a column with an adequate length and internal diameter (i.d.) are all means of improving resolution in HPLC. Most times, adjusting mobile phase ratio and changing flow regimes (isocratic vs gradient) will accomplish the desired resolution. Since the resolution is proportional to the square root of column length and the square of column i.d., the resolution can be improved to some extent with the use of a longer column or a column with a larger i.d. Increasing column length will increase resolution, but will also proportionally increase the retention time leading to a slower analysis.

12) **In selecting GC column temperature, what factors need to be considered?**

    Column temperature is the easiest and most efficient way to optimize separation in GC. The control of column temperature within ±0.5 °C is essential. Raising the column temperature speeds both the elution (higher vapor pressure) and the rate of approach to equilibrium between the mobile and stationary phases. In GC, the change in temperature as the separation proceeds is called a "temperature gradient," which can range from the ambient to 360°C. Higher column temperature should be avoided, since it will cause column bleeding where the stationary phase itself can be vaporized and/or decomposed and the material is then passed along the column and eluted. In addition, the temperature of the column should be lower than that of the downstream detector in GC. This is to prevent condensation of the sample and/or liquid phase in the detector.

13) **To separate very polar components by HPLC, would you choose a very polar or nonpolar material for the stationary phase? Which component (most polar to least polar) will be eluted first?**

In commonly used reverse-phase HPLC (RP-HPLC) systems, stationary phases such as C_8 or C_{18} columns are relatively nonpolar and the mobile phases are polar (such as the mixture of water and acetonitrile or methanol). Hence, a more polar compound will be eluted first before other less polar compounds. To separate very polar components by HPLC, a more polar stationary phase and/or more hydrophobic (less water) mobile phase is preferred. If the separation is not accomplishable in RP-HPLC, then try the normal phase HPLC using a column with a very polar stationary phase and a nonpolar mobile phase (such as hexane).

14) **Sort the stationary phase by the increasing hydrophobicity: (a) –CN, (b) –(OH)$_2$, (c) –(CH$_2$)$_7$CH$_3$, (d) –NH$_2$.**

(b) < (d) < (a) < (c)

15) **Define the following terms: separation factor, retention factor, resolution, plate number, and plate height. Which one best defines the column efficiency?**

Separation factor (α) is defined as the ratio of distribution constants (K) of two solutes. Retention factor (k) is defined as the ratio of the amount of a solute in the stationary phase to the amount in the mobile phase. Resolution (R) is the measure of column's ability to separate two peaks. Plate number (N) is the measure of the separation performance of column efficiency, and a plate height can be given as column length (L) divided by the plate number (N).

Column efficiency can be described by the plate number. A good column will have a large N and small H. In theory (Eq. 11.9), it is important to realize that the resolution can be optimized with combined values of α, N, and k. The separation factor (α) has the *largest* effect. If a separation needs to be improved, it is well worth the effort of increasing α, although it is impossible to give a general proposal on how to do this. If the plate number (N) is increased, the effect is only by the factor $\sqrt{N}$. This means that if the column length is doubled, the resolution will improve only $\sqrt{2} = 1.4$. Increasing the retention factor (k) only has a notable influence on resolution if k is small to start with.

16) **Explain why is it always a compromise between resolution and speed of analysis during a chromatographic analysis.**

Under given chromatographic conditions, a higher resolution can be achieved typically at the expense of longer run time, hence the slower analytical speed. For example, a longer column will improve resolution but will result in a longer sample run time.

17) **Why is a capillary column more efficient than a packed column in separation?**

One major factor is that all capillary columns can be made much longer than packed columns (10–100 m vs 1–5 m). Such long lengths permit much more efficient separations.

18) **Explain (a) how column length affects resolution. (b) Why HPLC column cannot be made very long for a better resolution?**

Increasing column length is one of the effective ways to improve resolution. A longer column will have a better chance of separating different compounds. In theory, if the column length is doubled, the resolution will improve by $\sqrt{2} = 1.4$. Unfortunately, a very long column cannot be employed in HPLC. This is because an extremely high pressure will be required to let mobile phase flow through a tightly packed HPLC column. On the contrary, capillary columns in GC can be made very long (e.g., several hundred meters long for GC columns vs centimeters in length for HPLC), since capillary columns are hollow tubes and the mobile phase flowing through is a gas rather than a liquid.

19) **Describe the effects of increased column length on: (a) column efficiency, (b) plate number, (c) time of analysis, (d) peak width.**

The plate number (*N*) is a measure of the separation performance of a column, or the column efficiency. Its relation to column length (*L*) and a theoretical plate (*H*) is as follows: $N = L/H$, where one theoretical plate with height *H* represents a single equilibrium step. A column with more theoretical plates (*N*) will have a greater resolving power. Increasing column length clearly will increase the plate number and the column efficiency. As a result, increased column length will also increase resolution, but it will also proportionally increase the retention times, resulting in slow analysis or wider peaks of all analytes.

20) **Explain the difference between the following: (a) reverse-phase HPLC and normal phase HPLC, (b) isocratic flow and gradient flow.**

a) The reverse-phase HPLC (RP-HPLC) has nonpolar stationary phase and relatively polar solvent mobile phase such as water, methanol, acetonitrile, or a mixture of water with one of the organic solvents. In RP-HPLC, the most polar component elutes first, and increasing mobile phase polarity increases the elution time. The normal phase HPLC (NP-HPLC) has the highly polar stationary phase and relatively nonpolar solvent mobile phase such as hexane, methylene chloride, or chloroform. In NP-HPLC, the polar stationary phase and nonpolar mobile phase favor the retention of polar compounds so that the least polar component elutes first. The least polar component is also the most soluble component in the hydrophobic mobile phase solvent. Increase in the polarity of the mobile phase will decrease the elution time.

b) The isocratic flow regime in HPLC has a constant solvent or mixed solvents composition over run time, whereas the gradient flow regime varies its solvent or mixed solvent composition over the run time.

21) **What will be the effect on retention time (i.e., increased or decreased) if the polarity of mobile phase is (a) decreased in normal phase HPLC; or (b) increased in reverse-phase HPLC.**

a) If the polarity of mobile phase is decreased in a normal phase HPLC, retention time will increase. (Hydrophilic compounds are the analytes of interest in NP-HPLC.)

b) If the polarity of mobile phase is increased in reverse-phase HPLC, retention will increase. (Hydrophobic compounds are the analytes of interest in RP-HPLC.)

22) **You are to separate three chemicals (A, B, C) on a reverse-phase HPLC column with a UV detector. You first tried isocratic flow with a mobile phase of 10% acetonitrile and 90% water at a flow rate of 1 mL/min. The retention times at this condition are 2.5, 5.5, and 5.9 min for chemical A, B, and C, respectively. Which chemical is most polar? Since the retention times for chemical B and C are so close and peaks are little overlapped, what would you do to better separate these two compounds? Why?**

For reverse-phase HPLC, the most polar compound elutes first. Therefore, compound A is the most polar among these chemicals. To better separate compounds B and C, we could increase the mobile phase polarity, because increasing the mobile phase polarity (e.g., more water) will also increase the elution time for these chemicals and make B and C a little more spread out. Alternatively, we could also try the gradient flow.

23) **Using the properties of solvents given in Table 11.3, answer the following: (a) why acetone is not used for HPLC with UV detector? (b) Why isopropanol often produces high pressure and cannot be used at higher percentage in mobile phase? (c) Which one is most likely preferred in RP-HPLC with UV detector: methanol, hexane, isopropanol, or acetone?**

a) Acetone is not typically used in HPLC because it has the highest UV cutoff wavelength of 330 nm among all the common solvents listed. This limits its use in HPLC as a mobile phase or as a diluting solvent.

b) Isopropanol (isopropyl alcohol) has the highest viscosity (2.4 centipoise) among all the solvents listed. High flow rates will result in over-pressure of the HPLC system.

c) Both methanol and acetonitrile are commonly used in RP-HPLC because of their low UV cutoff values and the intermediate polarity. In an increasing order, the polarity values are pentane = 0.1 (least polar), methanol = 5.1, acetonitrile = 5.8, water = 10.2 (most polar). Hexane is immiscible with water; hence it is not commonly used in RP-HPLC. In some cases, isopropanol is used in RP-HPLC. Acetone should be avoided in UV-based HPLC because of its high UV cutoff value.

24) **For a reverse-phase chromatography, (a) which of the following, 1:1 water:methanol; 1:1 water:acetone; 1:1 water:THF, has the highest strength in eluting a hydrophobic contaminant from a reverse-phase column (use solvent properties in Table 11.3); (b) which of the following stationary phase, unbonded silica, cyano, C_3, C_8, C_{18}, has the strongest retention for nonpolar, nonionic compounds?**

a) Methanol (CH_3OH) is slightly more polar than acetone [$(CH_3)_2C=O$]. Hence in comparison, a hydrophobic contaminant will be eluted a little easily with 1:1 water acetone. Tetrahydrofuran (THF, C_4H_8O) has the lowest polarity among all these three solvents (polarity = 4.0), so it has the highest strength in eluting a hydrophobic contaminant in RP-HPLC.

b) For hydrophobic compounds in RP-HPLC, C_{18} column has the longest (18) carbon chain stationary phase; it will be the strongest in retaining nonpolar, nonionic compounds.

25) **Define UV-cutoff, refractive index, and polarity index. Why are they important in HPLC?**

UV-cutoff is a property related to a solvent's transparency of UV radiation. It is defined as the wavelength at which the absorbance of the solvent vs air in 1-cm matching cells is equal to unity. In practice, the analytical wavelength (λ) of a UV detector should be at least 20 nm above the UV-cutoff value of the solvent.

Refractive index (RI) is a solvent's property in changing the speed of light. It is computed as the ratio of the speed of light in a vacuum to the speed of light through the solvent. The higher the RI, the slower the speed of light through the solvent. Solvent RI affects the sensitivity of RI detection for a particular sample. For a better sensitivity, the RIs of both the solvent and the sample should be as great as possible.

Polarity index (P) is a numerical measure of relative polarity of various solvents. Polarity index (P) ranges from $P = 0$ for a nonpolar solvent like pentane to $P = 10.2$ for the most polar solvent, water. Solvent strength increases as solvent polarity decreases in reverse-phase HPLC.

26) **Draw a schematic diagram showing the essential components of (a) HPLC, and (b) GC-FID.**

a) Students' answers will vary. Refer to Fig. 11.11 for a schematic of HPLC. Major components of a typical HPLC include: (1) solvent reservoirs and a degassing unit; (2) a solvent pump; (3) a sample injection system; (4) a column (guard column and analytical column); (5) a detector; and (6) a data-acquisition system.

b) Students' answers will vary. Refer to Fig. 11.10 for a schematic of GC-FID. Major components of a typical GC-FID include: (1) carrier gas supply (usually helium); (2) flow control for helium as a carrier gas and air-hydrogen for the flame; (3) sample introduction and splitter; (4) separation column; (5) temperature control zones (ovens); (6) detector (FID); and (7) a data-acquisition system.

27) **Describe the general strategy for keeping different temperatures in different zones of GC (i.e., injector, column oven, and detector). Why this is important?**

There are three temperature zones in a regular GC. (1) For the injection port, a general rule is to have the temperature at 50°C higher than the boiling point of the sample. This temperature should be high enough to vaporize the sample rapidly, but low enough to not thermally decompose the analytes. (2) In the column (oven) temperature zone, temperature is programmed to optimize compounds' separation. The control of column temperature within ±0.5°C is essential. Raising the column temperature speeds both the elution (higher vapor pressure) and the rate of approach to equilibrium between the mobile and stationary phases. In GC, the change in temperature as the separation proceeds is called a "temperature gradient," which can range from the ambient to 360°C. Higher column temperature should be avoided, since it will cause column bleeding where the stationary phase itself can be vaporized and/or decomposed and the material is then passed along the column and eluted. (3) The detector temperature depends on the type of detector used in the GC. As a general rule, however, detector temperature must be higher than the column eluent to prevent condensation of the sample and/or liquid phase.

28) **In ion chromatography commonly employed in ion analysis, explain (a) the separation principle, (b) why eluent is important, and (b) why suppressor column is essential.**

a) The separation in IC is based on ion exchange of ionic species instead of adsorption or partitioning.

b) Rather than using organic solvents of various polarities in HPLC, IC uses acid/base or salt buffer solutions as an eluent. The type and strength of the eluent are important so that various ionic analytes will have different affinities.

c) The suppressor column is essential because it reduces the background conductivity of the eluent to a low or negligible level.

29) **For anion analysis, what type (anionic or cationic) of exchange resin is used in analytical column? what type of exchange resin is used in suppressor column? What if cationic analysis is performed?**

For *anion* analysis, the analytical column is packed with a positively charged cationic exchange resin called anion-exchange column. The suppression column is placed downstream of the analytical column. This suppressed column is a large capacity anionic exchange resin used to retain cations and at the same time convert anions into their corresponding acids. The suppressed column is used to reduce the background conductivity of the eluent to a low or negligible level.

For *cation* analysis, the analytical column is packed with a negatively charged anionic exchange resin called cation-exchange column. The suppression column is placed downstream of the analytical column. This suppressed column is a large capacity cation-exchange resin used to retain anions and at the same time convert cations into their corresponding metal hydroxides (MOH). The suppressed column is used to reduce the background conductivity of the eluent to a low or negligible level.

30) **Describe the principles of measuring F^-, Cl^-, NO_3^-, and SO_4^{2-} in water by anionic IC with Na_2CO_3–$NaHCO_3$ as the eluting solvent.**

Anionic IC is a specialized type of system similar to an HPLC. In this case, there is a positively charged ion-exchange resin that will attract and hold the anions (F^-, Cl^-, NO_3^- and SO_4^{2-}) that are in the solution of interest. Because of the different affinities with the exchange resin, these anions will be eluted out into the suppressor column at various times, which in this case would have a very large capacity for holding cations. The anions would be turned to acidic forms and flow out of the suppressed column, and be measured by electrical conductivity.

31) **Name three most commonly used detectors in GC, and one most commonly used detector in (a) HPLC and (b) IC.**

GC has versatile choices of detectors. Three common GC detectors are flame ionization detector (FID), electron capture detector (ECD), and thermal conductivity detector (TCD). In HPLC, UV detector is the most commonly used one. There are two types of UV detectors: variable wavelength detector (VWD) and diode array detector (DAD). The VWD is unable to provide a UV spectrum, whereas a DAD can provide UV spectrum for the analytes of interest. The most commonly used detector in IC is the conductivity detector.

32) **Briefly explain the following components in GC: (a) septum; (b) inlet; (c) split/split-less injector; (d) liner; (e) on-column injection.**

a) Septum is a polymeric silicon through which the sample is directly introduced (injected through a syringe) into the heated region called inlet.

b) Inlet is the device that accepts the sample and transfers it to a GC column. In the GC inlet, liquid samples are vaporized and carried to the column.

c) Split/splitless injector is an option to control the amount of samples swept into the column and detector. In splitless mode, all vaporized samples enter into the column whereas in the split mode, sample quantity is to be reduced by the controlling value at a certain split ratio.

33) **Briefly explain the purpose of the following components in HPLC: (a) in-line filter; (b) guard column; (c) degassing unit.**

a) An in-line filter is replaceable filter for removing particulates generated by mobile phase or piston seals. This is different from the inlet filter placed inside each solvent container, which is used to protect the system from particulate matter or residual salt from entering.

b) A guard column is a very short replicate of analytical column with regard to the packing material. It is used to protect the analytical column from clogging with *sample-born* materials. Note that the inline filter only removes particulates from *mobile phase*.

c) A degassing unit is used to remove any gas from the HPLC system. Gas is detrimental to the high-pressure liquid chromatography system. A degassing unit can be a sintered glass connected to a helium gas cylinder, or an online vacuum degassing system in which solvent passes through a thin-walled porous polymer tube in a vacuum chamber.

34) **Explain the following: (a) Why is ECD highly sensitive to chlorinated pesticides such as DDT? (b) Why is PCE more sensitive than TCE in ECD? (c) Why is C_{12} hydrocarbon more sensitive than C_6 hydrocarbon in FID?**

a) DDT has four Cl atoms in its molecule (see structure below). A chlorine atom is highly electronegative because of its tendency to gain an additional electron to fully fill the outmost electronic shell with seven electrons. The ability to capture electrons is the base for its strong signal in ECD.

b) The response of ECD to chlorinated hydrocarbons increases by about 10 with each additional Cl atom. Hence, the sensitivity is in an increasing order of DCE < TCE < PCE (see structure below).

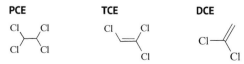

PCE   TCE   DCE

c) FID responds to the number of carbon atoms entering the detector per unit of time. The amount of ions produced is roughly proportional to the number of reduced carbon atoms present in the flame. As a result, C_{12} hydrocarbon is more sensitive than C_6 hydrocarbon in FID.

35) **Describe (a) what will chemical functionalities contribute to fluorescence and can be measured by a fluorescence detector? (b) how does halogen substitution affect a compound's ability to fluoresce.**

a) Fluorescence detectors respond only to fluorescent compounds. For the majority of fluorescent compounds, the radiation is produced by either an $n \rightarrow \pi^*$ or a $\pi \rightarrow \pi^*$ transition. The most intense and the most useful fluorescence is found in compounds containing aromatic functional groups with low-energy $\pi \rightarrow \pi^*$ transition levels. This explains why HPLC-fluorescent detector finds wide applications for the analysis of polycyclic aromatic hydrocarbons (PAHs) of environmental concern. The intensity of fluorescence increases as the number of fused benzene rings increases.

b) Halogen substitution generally results in a substantial decrease in fluorescence, and the substitution of a carboxylic acid (COOH) or carbonyl group (C=O) on an aromatic structure may inhibit fluorescence.

36) **Compare HPLC UV detectors with HPLC fluorescence detectors in terms of sensitivity. Why for the latter, is a 90^0 right angle generally used to measure the fluorescence intensity?**

The emission-based fluorescence detector (FLD) is very different from absorption-based UV detector. FLD is more selective and sensitive than UV detectors due to the noise reduction.

For fluorescence detectors, two wavelengths are concerned rather than the single wavelength used in absorption-based UV detectors. The detector employs an excitation source (typically a UV lamp), which emits UV radiation at a range of wavelengths. One or more filters or a grating monochromator in more sophisticated instruments are used to acquire the needed exciting beam ($\lambda_{excitation}$). The emitted light (fluorescence) is most conveniently measured at a 90^0 angle to the exciting light beam at a wavelength of $\lambda_{emission}$. This right angle is important, because at other angles, increasing light scattering for the solution and the cell walls may result in large errors during fluorescence measurement.

37) **Give examples of universal and selective detectors for both GC and HPLC.**

Examples of common universal detectors used in GC are TCD, FID (universal for hydrocarbons), and MS. Examples of common universal detectors used in HPLC are refractive index detector and MS.

38) **Given the following five GC detectors and the five VOCs or SVOCs, find the best one-to-one matching between these two. Available detectors are: (1) TCD, (2) FID, (3) ECD, (4) NPD, (5) FPD. Chemicals to be measured are: (a) merphos (CAS 150-50-5; (n-CH$_3$CH$_2$CH$_2$CH$_2$)$_3$P; a pesticide); (b) methyl methanesulfonate (CAS 66-27-3; CH$_3$SO$_2$OCH$_3$); (c) hexachlorobutadiene (CAS 87-68-3; Cl$_2$C=CClCCl=CCl$_2$); (d) ethylbenzene (CAS 100-41-4; (C$_6$H$_5$)C$_2$H$_5$); (e) mercury vapor.**

1) TCD for (e) mercury vapor: TCD is a universal detector, which is based on changes in heat conductivity. It would respond to all the compounds listed. None of the other detectors listed, however, will respond to mercury vapor.

2) FID for (d) ethylbenzene: FID is able to detect all hydrocarbons (a) through (d). Given the constraint of one-to-one match, however, the FID is the only option.

3) ECD for (c) hexachlorobutadiene: This is the only halogenated compound, so ECD should be the only detector of selection.

4) NPD for (a) merphos: NPD is a selective detector for N- and P-containing organic compounds.

5) FPD for (b) methyl methanesulfonate: FPD is a selective detector for S- and P-containing compounds. FPD would also be a useful detector for (a) merphos, because it is a P-containing compound.

39) **For the following compounds and the sample matrix, which chromatographic instrument (GC, HPLC, IC) and detector would you recommend? Suggest an EPA standard method for each chemical as well: (a) trichloroethene (volatile) in air; (b) PAHs in solid wastes; (c) BETX in soil; (d) acid rain composition (SO_4^{2-}, NO_3^-, and Cl^-); (e) species of chromium (Cr^{6+}) in water.**

a) GC is an option for volatile trichloroethene. The US EPA ambient air test methods for air toxins (TO 1 to TO-17) should be consulted, such as TO-1 to TO-4, TO-7, TO-9, TO-10, TO-12 to TO-17.

b) Since PAHs are semivolatile compounds, both GC and HPLC can be employed. SW-486's 8000 series methods will be a good starting point for PAHs analysis in solid wastes. One such method is SW-846's Method 8310 used to determine the concentration of certain PAHs in groundwater and wastes using HPLC.

c) GC would be the method of choice because of the high volatility of BTEX compounds. Since the sample is in a solid matrix, Method 8021 in SW-846 would be the method of choice. It is used to determine volatile organic compounds in a variety of solid waste matrices. This method is applicable to nearly all types of samples, regardless of water content, including groundwater, aqueous sludges, caustic liquors, acid liquors, waste solvents, oily wastes, mousses, tars, fibrous wastes, polymeric emulsions, filter cakes, spent carbons, spent catalysts, soils, and sediments.

d) For multiple anions (SO_4^{2-}, NO_3^-, and Cl^-) analysis, ion chromatography is the only option. A general search by matrix, source, and instrument in NEMI database (http://www.nemi.gov) gives two EPA methods: 300.0 (Inorganic Anions by Ion Chromatography, EPA-NERL) and 300.1 (Anions in Water by IC, EPA-OGDW/TSC).

e) Various species of chromium in water can be analyzed by conductivity-based ion chromatography. A search in NEMI database gives the following EPA methods:

Method number	Method descriptive name	Source
1636	Hexavalent chromium by ion chromatography	EPA-EAD
218.6	Hexavalent chromium, dissolved, in water by IC	EPA-NERL
218.6	Hexavalent chromium in water by ion chromatography	EPA-ORD/EPA-OST
7199	Chromium in water by ion chromatography	EPA-OSW

40) **Why are HPLC and GC mostly complementary with regard to the chemicals each can analyze? If a chemical can be analyzed by both methods, in what case, one is preferred over the other? How and why?**

The compounds that can be analyzed *directly* by GC are limited to gases, volatile organic compounds (VOCs), and semivolatile compounds (SVOCs). However, it is estimated that approximately 85% of known organic compounds are not sufficiently volatile or thermally stable enough to be separated and analyzed by GC. In this regard, HPLC is complementary to GC. HPLC should have a greater potential in analyzing organic compounds of semivolatile or nonvolatile nature.

One of the good features of HPLC is its ability to directly inject aqueous samples. Hence, HPLC can be a method of choice with regard to the time-saving for sample preparation. However, GC has many versatile detectors, and in many cases, derivatization enhanced this technique for some nonvolatile compounds. HPLC, on the other hand, is still less affordable and involves more troubleshooting.

41) **List the desirable characteristics of (a) GC and (b) HPLC.**

In addition to the types of compounds being analyzed by GC and HPLC, a general comparison of advantages and disadvantages can be made between these two. Advantages of GC include: (a) low cost, fast analysis, and ease of operation; (b) more sensitive and higher resolution when capillary columns are used; (c) a variety of columns (stationary phases) and variety of general and selective detectors to choose from, providing analytical flexibility; (d) readily interfaced with mass spectrometer for structural confirmation. Advantages of HPLC include: (a) direct analysis of aqueous sample without tedious sample preparation; (b) mobile phases of various polarities provide versatility. Compared to GC, HPLCs are generally more expensive, less sensitive, and slower. Besides, HPLC instruments are not readily interfaced with mass spectrometer, and operationally, more problematic with hardware.

42) **Explain the common causes of HPLC having: (a) low pressure; (b) high pressure; and (c) pressure fluctuation.**

a) Low pressure could be air bubbles, mis-set backpressure regulator, leaks in unions and fittings, broken piston, or major leaks around a piston seal.

b) High pressure could be due to clogged in-line filter, clogged guard column/analytical column, or clogged connected tubing. High pressure can also be the inadequate large flow rate, immiscible solvents, or use of large % of high viscosity mobile phase such as isopropanol.

c) Fluctuated pressure may be due to damaged pump valves and/or worn seal, air bubble in the pump head, or partially clogged pump inlet filter.

43) **Explain the common causes for both GC and HPLC, regarding: (a) fronting peak; (b) tailing peak; (c) ghost peak; and (d) split peak.**

a) Fronting peaks are caused by sample overloading.

b) Tailing peaks are caused by severe column contamination.

c) Ghost peaks are caused by contaminated sample or mobile phase, carry-over from previous analysis, or perhaps inlet/septum bleed.

d) Split peaks are usually caused by a bad injection technique.

44) **Explain the common causes that result in the following baseline problems: (a) noisy and spikes; (b) drifting; (c) a gradual rise at the end of high-temperature range (GC only).**

a) In GC, the common causes for a noisy baseline could be contaminated gases, inlet, column and/or detector, septum degradation, or leaks when using an MS, ECD, or TCD detector. In HPLC, noisy baseline commonly occurs with dirty samples, dirty detector cell, trapped air bubbles, or not fully equilibrated mobile phase.

b) In GC, when erratic baseline (wander and drift) appears, it is a sign of contaminated gas and column, incompletely conditioned column, change in gas flow rate, or large leaks at septum during injection and for a short time thereafter. In HPLC, drifting baseline is a sign of not fully equilibrated system, change in mobile phase due to continuous sparging, or change in temperature.

c) When a rise in baseline occurs as column approaches high temperature, this is typical of column bleeding when maximum allowable column temperature is approached or exceeded. Here "bleeding" refers to the breakdown of GC column stationary phase at a high temperature. The column needs to be reconditioned, trimmed, or replaced.

45) **What causes GC column bleeding and how to prevent it?**

Column bleeding in GC is caused by contaminated gas and column, incompletely conditioned column, change in gas flow rate, or leaks in the septum during injection and shortly thereafter. Most times, the bleeding is caused by the breakdown of GC column stationary phase at a high temperature. To combat column bleeding, the column should be re-conditioned, trimmed or replaced.

46) **Suppose you are involved in a GC-MS lab, you injected a mixed calibration standard solution that included six benzene series compounds (benzene, toluene, *o*-xylene, *m*-xylene, *p*-xylene, ethylbenzene) and four chlorobenzene series compounds (chlorobenzene, 1,2-dichlorobenzene, 1,3-dichlorobenzene, 1,4-dichlorobenzene). However, somehow you could not find all 10 peaks but observed a total of only seven peaks in the total ion chromatogram (TIC). The GC-MS library search also indicated that benzene is apparently not shown in the chromatogram. In addition to seven major peaks, you also noticed a small peak that does not match any of the chemicals listed above. For each of these three observations, explain what the reason(s) could be, and if there are any possible way to solve the problem and/or improve the method.**

Students' answers may vary. (1) The missing three peaks could be the loss of the most volatile compounds such as benzene, or perhaps the co-elution of several isomers. (2) Benzene peak is missing mostly because the temperature is set too high for the inlet. It could be eluted during the solvent delay period, during which period GC-MS does not collect any mass spectrum signals. (3) Ghost peak(s) are mostly caused by contaminated samples or carrier gas, carryover from previous analysis, or perhaps inlet/septum bleed. Apparently in this case, the GC column should first be baked, and any cross-contamination from carrier gas or syringe should be checked. After a clean baseline is obtained, the GC-MS method parameters (inlet temperature, oven temperature, etc.) need to be refined so that a better resolution can be obtained to separate all 10 compounds. It is possible that some of the isomers may still co-elute, so another GC column should be sought to achieve the best resolution. Contact the column providers since many of these common chemicals may already have the application notes as to what column is appropriate and what analytical conditions should be followed.

Problems

1) **The following data apply to a column liquid chromatography: length of packing = 24.7 cm; flow rate = 0.313 mL/min. A chromatogram of a mixture of Chemical A and B provided the following data:**

	Retention time, min	Width of peak base (*w*), min
Nonretained solvent	3.1	–
Chemical A	13.3	1.07
Chemical B	14.1	1.16

Calculate the following and make comments regarding the separation:

a) **The number of plates (*N*) from each peak (Chemical A and B)**
b) **The plate height (*H*) for the column based on the average number of plates**
c) **The retention factor (*k*) for each peak**
d) **The resolution between Chemical A and B**

e) **The separation factor (α) between Chemical A and B**

f) **The length of column necessary to give a resolution of 1.5**

a) $N_A = 16\left(\dfrac{t_{rA}}{w_A}\right)^2 = 16\left(\dfrac{13.3}{1.07}\right)^2 = 2472$; and $N_B = 16\left(\dfrac{t_{rB}}{w_B}\right)^2 = 16\left(\dfrac{14.1}{1.16}\right)^2 = 2364$

b) N (average) $= (2472 + 2364)/2 = 2418$

   Plate height (H) $= L/N = 24.7$ cm$/2418 = 0.010$ cm

c) Retention factor (k):

   $k_A = \dfrac{t_{rA} - t_m}{t_m} = \dfrac{13.3 - 3.1}{3.1} = 3.29$; and $k_B = \dfrac{t_{rB} - t_m}{t_m} = \dfrac{14.1 - 3.1}{3.1} = 3.55$

d) Resolution (R):

   $R = \dfrac{t_{rB} - t_{rA}}{(w_A + w_B)/2} = \dfrac{14.1 - 13.3}{(1.07 + 1.16)/2} = 0.72$

e) Separation factor (α):

   $\alpha = \dfrac{k_B}{k_A} = \dfrac{t_{rB} - t_m}{t_{rA} - t_m} = \dfrac{3.55}{3.29} = \dfrac{14.1 - 3.1}{13.3 - 3.1} = 1.06$

f) Using the following equation between plate height and resolution:

   $$H_2 = H_1 \times \dfrac{(R_2)^2}{(R_1)^2} = 24.7 \text{ cm} \times \dfrac{(1.5)^2}{(0.72)^2} = 107 \text{ cm}$$

   The resolution (R) of 0.72 is too low so these two peaks have a much overlap. Since the retention factors (k) are in the good range of 2–10, the separation of these two compounds can be improved. Based on the calculated column length (H_2), however, this appears that changing it to a longer column alone is not a feasible option. Practically, an HPLC column cannot be this long (107 cm).

2) **Refer to Fig. 11.9, use a ruler to measure the retention times, and calculate N, H, k, α for Chemical A and B. Assume the time unit is in minutes.**

   One can use a ruler to measure the length in centimeters in Fig. 11.9, which is assumed to be proportional to the retention time in min.

   For Chemical A:

   $N = 16\left(\dfrac{t_r}{w}\right)^2 = 16\left(\dfrac{3.4}{0.5}\right)^2 = 740$

   $H = \dfrac{L}{N} = \dfrac{L}{740}$ (where L is the column length, not given in this problem)

   $k = \dfrac{t_r - t_m}{t_m} = \dfrac{3.4 - 0.25}{0.25} = 12.6$

   $\alpha = \dfrac{k_b}{k_a} = \dfrac{t_{r,b} - t_m}{t_{r,a} - t_m} = \dfrac{5.4 - 0.25}{3.4 - 0.25} = 1.63$

   For Chemical B:

   $N = 16\left(\dfrac{t_r}{w}\right)^2 = 16\left(\dfrac{5.4}{0.9}\right)^2 = 576$

   $H = \dfrac{L}{N} = \dfrac{L}{576}$

   $k = \dfrac{t_r - t_m}{t_m} = \dfrac{5.4 - 0.25}{0.25} = 20.6$

   $\alpha = \dfrac{k_b}{k_a} = \dfrac{t_{r,b} - t_m}{t_{r,a} - t_m} = \dfrac{5.4 - 0.25}{3.4 - 0.25} = 1.63$

3) **A chromatographic analysis of a pesticide 1,4'-DDT using a 2-m long packed GC column (2.0 m × 4 mm id glass) gives a peak with retention time of 14.80 min and a baseline width of 0.95 min. Calculate (a) the number of theoretical plates and (b) the average height of a theoretical plate in mm.**

$$N = 16\frac{t_r^2}{w^2} = 16\frac{14.8^2}{0.95^2} = 3883$$

$$H = \frac{L}{H} = \frac{2 \text{ m}}{3883} = 0.000515 \text{ m} = 0.515 \text{ mm} / \text{plate}$$

4) **For bromide and nitrate in the chromatogram (Fig. 11.20), calculate the number of theoretical plates and the average height of a theoretical plate. An example column used for the analysis of anions using ion chromatography is a 25-cm long Metrosep A Supp 7-250/4.0 (Lawal et al., 2010, *J. Chromatogr. Sci.*, 48:537–543).**

Because the relationship between elution time and distance is proportional, we can measure the retention time and peak width using a ruler. Using these values, the number of theoretical plates and height for each plate are:

For bromide:

$$N = 16\frac{t_r^2}{w^2} = 16\frac{3.95^2}{0.35^2} = 2038$$

$$H = \frac{L}{N} = \frac{0.25 \text{ m}}{2038} = 0.000123 \text{ m} / \text{plate} = 0.123 \text{ mm} / \text{plate}$$

For nitrate:

$$N = 16\frac{t_r^2}{w^2} = 16\frac{4.60^2}{0.50^2} = 1354$$

$$H = \frac{L}{N} = \frac{0.25 \text{ m}}{1354} = 0.000185 \text{ m} / \text{plate} = 0.185 \text{ mm} / \text{plate}$$

Students' measurements for t and w will depend on the relative size of computer monitors or printouts; however, the calculated values for N and H should be similar to the answers above.

Chapter 12

Questions

1) **Explain why electrochemical methods measure the activity rather than concentration.**
   The underlying equation for electrochemical methods is Nernst equation. For a redox reaction, a A + b B ⇌ c C + d D, the Nernst equation relates electrode potential (E) and activities (in {}) of chemical species as follows:

$$E = E^0 - \frac{RT}{nF}\ln\frac{\{C\}^c\{D\}^d}{\{A\}^a\{B\}^b}$$

   where E^0 = electrode potential in the standard state, R = ideal gas law constant (8.314 Joule $\times$ K^{-1} $\times$ mole^{-1}), T = temperature (K), n = number of electrons in the half cell-reaction, F = Faraday constant (9.65×10^4 Coulomb $\times$ mole^{-1}). It is important to note that the Nernst equation relates the potential to activities, i.e., the "effective" concentrations, rather than the molar concentrations. The activity $\{X\}$ is related to the concentration $[X]$ of species X as:

$$\{X\} = \gamma \times [X]$$

   where γ is the activity coefficient, which tends to be unity at low concentrations. Because of this, activities in the Nernst equation are frequently replaced by concentrations for simplification as follows:

$$E = E^0 - \frac{RT}{nF}\ln\frac{[C]^c[D]^d}{[A]^a[B]^b}$$

   The above approximation is valid for ion concentrations < 0.01 mole/L. Otherwise, activity should be used in place of concentration.

2) **Given the E^0 values of the following two half-cell reactions:**

   $$\text{Zn} \rightarrow \text{Zn}^{2+} + 2\text{e}^- \quad E^0 = 0.763 \text{ volt}$$

   $$\text{Fe} \rightarrow \text{Fe}^{2+} + 2\text{e}^- \quad E^0 = 0.441 \text{ volt}$$

   a) **Write a balanced complete oxidation–reduction reaction.**

   b) **Explain whether the corrosion of an iron pipe (i.e., $\text{Fe} \rightarrow \text{Fe}^{2+}$) in the presence of Zn/$\text{Zn}^{2+}$ is thermodynamically possible or not.**

   c) **Explain whether or not Zn will protect the corrosion of iron pipe if metallic Zn is in direct contact with the iron pipe.**

Fundamentals of Environmental Sampling and Analysis, Second Edition. Chunlong Zhang.
© 2024 John Wiley & Sons, Inc. Published 2024 by John Wiley & Sons, Inc.
Companion Website: www.wiley.com/go/EnvironmentalSamplingandAnalysis2e

a) The complete oxidation–reduction reaction is obtained by reverting one of the half-cell reactions to cancel out the electrons:

$$Zn + Fe^{2+} \leftrightharpoons Zn^{2+} + Fe \quad E^0 = 0.763 - 0.441 = 0.322 \text{ volts}$$

Thus, two moles of electrons lost from elemental Zn are gained by Fe^{2+}.

b) The reaction as written in part (a) is spontaneous because of the positive value of the overall reaction potential of 0.322 volts. The corrosion reaction (i.e., $Fe \rightarrow Fe^{2+}$) is thermodynamically not favorable.

c) Elemental Zn will protect elemental iron from corrosion (i.e., $Fe \rightarrow Fe^{2+}$). Here, Zn is sacrificed because Zn itself is corroded (oxidized) based on the above spontaneous reaction from left to right.

3) **Construct a complete balanced oxidation–reduction reaction from each pair of two half-cell reactions:**

a) $H_2 \rightarrow 2H^+ + 2e^- \quad Fe^{2+} \rightarrow Fe^{3+} + e^-$

b) $1/4O_2(g) + H^+ + e^- \rightarrow \frac{1}{2} H_2O \quad H_2 \rightarrow 2H^+ + 2e^-$

c) $2IO_3^- + 12H^+ + 10e^- \rightarrow I_2 + 6H_2O \quad 2I^- = I_2 + 2e^-$

a) Reverting the second half-cell reaction and multiplying by 2 to cancel the electrons:

$$H_2 \rightarrow 2H^+ + 2e^-$$
$$2Fe^{3+} + 2e^- \rightarrow 2Fe^{2+}$$

-- (+

$$H_2 + 2Fe^{3+} \leftrightharpoons 2H^+ + 2Fe^{2+}$$

b) To cancel out the electrons, we need to multiply the first half-cell reaction by 2:

$$1/2O_2(g) + 2H^+ + 2e^- \rightarrow H_2O$$
$$H_2 \rightarrow 2H^+ + 2e^-$$

-- (+

$$1/2O_2(g) + 2H^+ + H_2 \leftrightharpoons H_2O + 2H^+$$

The above overall reaction can be further simplified by canceling out $2H^+$ on each side:

$$1/2O_2(g) + H_2 \leftrightharpoons H_2O$$

c) To cancel out the electrons, we need to multiply the second half-cell reaction by 5:

$$2IO_3^- + 12H^+ + 10e^- \rightarrow I_2 + 6H_2O$$
$$10I^- \rightarrow 5I_2 + 10e^-$$

-- (+

$$2IO_3^- + 12H^+ + 10I^- \leftrightharpoons I_2 + 6H_2O + 5I_2$$

Adding I_2 and $5I_2$, we have: $2IO_3^- + 12H^+ + 10I^- \leftrightharpoons 6H_2O + 6I_2$

This can be simplified by dividing both sides by 2. Then we have the overall reaction:

$$IO_3^- + 6H^+ + 5I^- \leftrightharpoons 3H_2O + 3I_2$$

4) **Given the electrochemical cell as shown in the figure on page 409, (a) identify the anode and cathode, (b) identify the negative and positive terminal, (c) determine the direction of electron flow. (d) Which one of the following is incorrect for the reaction: $H_2 + I_2 \rightleftarrows 2H^+ + 2I^-$? (i) Two moles of electrons are involved. (ii) The oxidation of H_2 takes place at the anode. (iii) Electrons move from a cathode to an anode.**

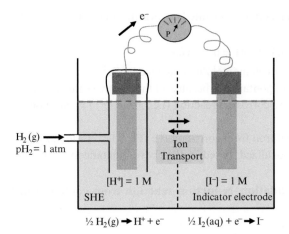

½ H$_2$(g) → H$^+$ + e$^-$ ½ I$_2$(aq) + e$^-$ → I$^-$

a) From the electron flow shown in the diagram, we know that one electron is lost from 1/2H$_2$ (oxidation at the anode). The 1/2I$_2$ gains this electron (reduction at the cathode). In other words, H$_2$ is oxidized to H$^+$, whereas I$_2$ is reduced to I$^-$.

b) H$_2$ is a reducing agent, whereas I$_2$ is an oxidizing agent. The reducing agent (i.e., H$_2$) provides electron in order to reduce the other chemical species (i.e., I$_2$).

c) Both (i) and (ii) are correct statements. (iii) is incorrect because electrons always flow from anode (oxidation half-cell) to cathode (reduction half-cell).

5) **For the two half-cell reactions below, which of these could serve as the anode if the combined reaction is to make (a) a galvanic cell, (b) an electrolytic cell?**

$$Ag^+\left(aq\right) + e^- \rightarrow Ag\left(s\right) \quad E^0 = +0.80 \text{ V}$$

$$Al^{3+}\left(aq\right) + 3e^- \rightarrow Al\left(s\right) \quad E^0 = -1.66 \text{ V}$$

Multiplying by 3 for the first half-cell reaction, and then reversing the second half-cell reaction and changing the sign of the E^0, we obtain:

$$3Ag^+\left(aq\right) + 3e^- \rightarrow 3Ag\left(s\right) \quad E^0 = +0.80 \text{ V}$$

$$Al\left(s\right) \rightarrow Al^{3+}\left(aq\right) + 3e^- \quad E^0 = +1.66 \text{ V}$$

Next, we add these two half-cell reactions to cancel out 3e$^-$ on both sides:

$$3Ag^+\left(aq\right) + Al\left(s\right) \rightarrow 3Ag\left(s\right) + Al^{3+}\left(aq\right) \quad E^0 = +2.46 \text{ V}$$

The above cell as written has a positive electrode potential of +2.46; it is thus a galvanic cell, meaning it generates electrical energy from chemical energy (a battery). The anode where the oxidation occurs is Al (s) → Al^{3+}. The opposite of the above reaction is:

$$3Ag\left(s\right) + Al^{3+}\left(aq\right) \rightarrow 3Ag^+\left(aq\right) + Al\left(s\right) \quad E^0 = -2.46 \text{ V}$$

The above reaction requires an external electric energy to drive the chemical reaction from left to right. It is an electrolytic cell. In this reaction, Ag (s) is oxidized to Ag$^+$ (aq), hence it is the anode.

6) **Write the standard cell notation for the cell shown in Problem 4.**

$$Pt\Big|H_2(1 \text{ atm}), H^+(1 \text{ M})\Big\|I_2, I^-\Big| Pt$$

Note that the conventions for shorthand notations require the reaction involving oxidation to be written on the left. Two platinum plates were assumed in the figure in Problem 4.

7) **Write the standard cell notation for the two cells shown in Fig. 12.5 employed for dissolved oxygen measurement.**

$$Ag(s) \,\big|\, AgCl(s), KCl(saturated) \,\big\|\, O_2, OH^- \,\big|\, Au$$

The membrane-based dissolved oxygen probe consists of a gold cathode where O_2 is reduced, and a silver anode where Ag is oxidized in a saturated KCl solution.

8) **What species is produced at the cathode in the cell noted below? Is this a galvanic cell or an electrolytic cell?**

$$Sn \,\big|\, Sn^{2+}(aq) \,\big\|\, Cl^- \,\big|\, Cl_2(g) \,|\, Pt$$

$$Sn^{2+}(aq) + 2e^- \rightarrow Sn(s) \quad E^0 = -0.14 \text{ V}$$

$$Cl_2(g) + 2e^- \rightarrow 2Cl^- \quad E^0 = +1.36 \text{ V}$$

By convention, the half-cell reaction involving oxidation is written on the left of the cell notation. Thus,

Anode(oxidation): $Sn(s) \rightarrow Sn^{2+}(aq) + 2e^- \quad E^0 = +0.14 \text{ V}$

Cathode(reduction): $Cl_2(g) + 2e^- \rightarrow 2Cl^- \quad E^0 = +1.36 \text{ V}$

The overall reaction is obtained by combining the above two half-cell reactions:

$$Sn(s) + Cl_2(g) \rightarrow Sn^{2+}(aq) + 2Cl^- \quad E^0 = +0.14 + (+1.36 \text{ V}) = +1.50 \text{ V}$$

Since the E^0 for the overall reaction is positive (+1.50), the above cell will spontaneously produce electrical energy and is therefore galvanic.

9) **The AA- and AAA-cell batteries are nonrechargeable with the following standard notation:**

$$Zn(s) \,\big|\, Zn(O)(s) \,\big|\, KOH(aq) \,\big\|\, KOH(aq) \,\big|\, Mn_2O_3(s) \,\big|\, MnO_2(s) \,|\, C(s)$$

where graphite powder (C) is mixed with MnO_2 to make a paste in the cathode to serve as electrode surface. Write the half-cell reactions in the anode and cathode, and the overall reaction.

Anode: $Zn(s) + 2OH^- \rightarrow ZnO(s) + H_2O + 2e^-$

Cathode: $2MnO_2(s) + H_2O + 2e^- \rightarrow Mn_2O_3(s) + 2OH^-$

-- (+

Overall: $2MnO_2(s) + Zn(s) \rightarrow Mn_2O_3(s) + ZnO(s)$

10) **The lithium-ion batteries have the following redox reactions during the discharging mode:**

$$CoO_2 + Li^+ + e^- \rightarrow LiCoO_2$$

$$LiC_6 \rightarrow C_6 + Li^+ + e^-$$

where LiO_6 is a graphite intercalation compound and C_6 is graphite. Identify the cathode and anode, develop the overall reaction, and write the standard cell notation.

Regardless of galvanic cells or electrolytic cells, the oxidation takes place in an anode, and the reduction takes place in a cathode:

Anode: $LiC_6 \rightarrow C_6 + Li^+ + e^-$

Cathode: $CoO_2 + Li^+ + e^- \rightarrow LiCoO_2$

-- (+

Overall reaction: $LiC_6 + CoO_2 \rightarrow C_6 + LiCoO_2$

Standard cell notation: $LiC_6 \,\big|\, C_6, Li^+ \,\big\|\, CoO_2, Li^+ \,\big|\, LiCoO_2$

11) **Why is TISAB added to the sample solution when ion selective electrode (ISE) such as fluoride electrode is used to measure the ion concentration?**

One major problem with ISEs is the interference from other ions. Other limitations include the effects of ionic strengths, and the potential formation of complex compounds that are not responsive to ISEs. Many of these problems, however, can be solved by adding a total ionic-strength adjustment buffer (TISAB). The commercially available TISAB for fluoride electrode, for example, is a mixture of an acetate buffer (pH 5.0–5.5), 1 M NaCl, and cyclohexylenedinitrilo tetraacetic acid (CDTA). The pH in this range will ensure a low OH^- concentration in the solutions while also avoiding the formation of appreciable HF, which is not responsive to the membrane. CDTA is a chelating agent, which can release F^- from possible complex compounds such as Al^{3+}, Fe^{3+}, and Si^{4+}.

12) **Explain the principles of ion-selective electrode used for the measurement of (a) pH and (b) nitrate.**

a) A key component for a pH meter is the electrode with a glass membrane that allows hydrogen ions (H^+) in the sample solution to enter the hydrated Si–O lattice structure of the glass membrane and exchange for singly charged cation Na^+. This creates an electrical potential across the membrane interface (referred to as the boundary potential) with respect to the internal Ag/AgCl reference electrode. The resulting overall cell potential with regard to the external reference cell can be derived as:

$$E_{cell} = \text{constant} + \frac{RT}{F} \ln a_{H^+} = \text{constant} - \frac{2.303RT}{F} pH_{unknown}$$

where a_{H^+} is the activity of H^+. The constant value is unknown and is not of practical interest for pH measurement, because standard pH buffers are usually used to calibrate the electrode. The slope factor ($2.303RT/F$) is temperature dependent; therefore, temperature compensation should be made during pH measurement.

Note that the above equation is strikingly similar to the Nernst equation (Eq. 12.1) used to describe the metallic electrodes. However, the source of potential in pH glass electrodes is totally different from metallic electrodes – one arises from *redox reaction*, whereas the pH electrode is due to a boundary potential as a result of *ion-exchange* reactions.

b) The nitrate electrode is a liquid membrane, which incorporates an ion exchanger along with a reference electrode. The membrane of commercial nitrate electrode is usually polyvinyl chloride (PVC). As the nitrate ion is a strongly hydrophilic anion, the anion exchanger is a strongly hydrophobic cation such as tetraalkylammonium salt. The working principle for nitrate-selective electrode is the same as other ISEs. Nitrate electrode serves as a transducer (or sensor) that converts the activity of nitrate dissolved in a solution into an electrical potential, which can be measured by a voltmeter or pH meter. The voltage is theoretically dependent on the logarithm of the nitrate activity, according to the Nernst equation.

13) **Is pH membrane electrode based on the redox reaction described by the Nernst equation? Why or why not.**

No, the pH membrane electrode is not based on the redox reaction like most ISEs. The underlying equation for pH measurement is not the Nernst equation, but an equation (Eq.12.8) very similar to Nernst equation:

$$E_{cell} = \text{constant} + \frac{RT}{F} \ln a_{H^+} = \text{constant} - \frac{2.303RT}{F} pH_{unknown}$$

where a_{H^+} is the activity of H^+. The constant value is unknown and is not of practical interest for pH measurement, because standard pH buffers are usually used to calibrate the electrode. The slope factor ($2.303RT/F$) is temperature dependent. The source of potential in pH glass electrodes is totally different from metallic electrodes – one arises from *redox reaction*, whereas the pH electrode is due to a boundary potential as a result of *ion-exchange* reactions.

14) **Describe the precautions for the use and maintenance of glass pH electrodes.**

pH meter is perhaps one of the most poorly understood and maintained apparatus in many labs. Below is a summary of precautions for the use and maintenance of glass pH electrodes:

- The pH probe should be fully hydrated prior to use by soaking the membrane in water for 24–48 h. Never let glass electrode dry out.
- The pH meter should be calibrated on a frequent basis, because many factors can cause errors, including stains within the membrane, mechanical or chemical attack of the external surface. Handle glass electrode with extreme care.
- The pH reading may be sluggish when the pH of a dilute or an unbuffered solution of near neutrality is measured. The solution should be stirred, and a sufficient time is required to obtain a stable pH reading.
- Like other sensor devices, general-purpose pH electrodes perform well in certain ranges. Solutions of extreme pH values (pH < 0.5 and pH > 10) will result in non-Nernstian behavior, which is termed alkaline error and acid error, respectively. Use a lithium glass electrode or a full-range electrode (0–14) instead, if the pH of very acidic or basic solutions is measured.
- Make sure to store a pH electrode in its wetting cap containing electrode fill solution (3 M KCl; purchased or prepared by dissolving 22.37 g KCl into 100 mL DI water). Do not store electrode in DI water.

15) **Give a list of common ions of environmental significance that can be analyzed by ion-selective electrodes.**

The table below is a summary of the common anions and cations that can be analyzed by ISEs of various types. They are grouped by the types of membranes.

Membrane	Analyte ions
Glass	$H^{+\,(1,\,2)}$, Ag^+, K^+, Li^+, Na^+, NH^{4+}
Solid phase	$Br^{-\,(1)}$, $Cl^{-\,(1,\,2)}$, $CN^{-\,(1,\,2)}$, $F^{-\,(1,\,2)}$, I^-, SCN^-, $S^{2-\,(1,\,2)}$, Ag^+, Cd^{2+}, Cu^{2+}, Pb^{2+}
Liquid phase	BF_4^-, ClO_4^-, $NO_3^{-\,(1,\,2)}$, Ca^{2+}, $K^{+\,(2)}$, hardness ($Ca^{2+} + Mg^{2+}$)

*Standard methods are available from (1) US EPA and (2) APHA/AWWA/WEF.

A note of the environmental applications is fluoride electrode and nitrate electrode. Fluoride electrode can measure fluoride as low as 1 µg/L level. Nitrate electrode can measure concentration down to 2 mg/L. It compares favorably with its colorimetric method because of its simplicity and its ability for *in-situ* measurement.

16) **Three electrochemical titration methods are discussed in this chapter: potentiometric, coulometric, and amperometric titration. (a) Describe the fundamental differences of the principles, and (b) give examples of environmental measurements.**

a) Principles:

Potentiometric titration is a volumetric method in which the potential (E) between two electrodes (reference and indicator electrodes) is measured as a function of the added reagent volume (V). The indicator electrode can be selected to respond to either a reactant or a product. A typical plot of E vs V has a characteristic sigmoid (S-shaped) curve. The part of curve that has the maximum change marks the equivalent point of titration. This point can also be determined by the slope of the curve (i.e., the first derivative, $\Delta E/\Delta V$) vs *V* plot.

Coulometric titration uses a constant current system to perform the reaction. The only measurement required in these systems is the time it takes to complete the electrolysis. The product of this time and the current is then used to determine the total amount of electricity used. The endpoint of the titration can be determined analytically by using an indicator that is placed in the sample and signals when the system reaches equilibrium. Alternatively, the endpoint can be determined from data provided by potentiometric, amperometric, or conductance measurements. This is similar to regular chemical titrations.

Amperometric titration refers to a class of titrations in which the equivalence point is determined through measurement of the electric current produced by the titration reaction. The output of the amperometric titration, unlike potential titration, is a plot of current vs titrant volume. The abrupt change in current is used as the endpoint.

b) Examples of environmental measurements:

Potentiometric titrations can be used not only for redox reaction but also for acid–base, precipitation, and complexation. In environmental measurement, potentiometric titrations are preferred for the determination of acidity and alkalinity.

Coulometric titrations can be used for the neutralization of acids, which contain H^+ ions, by producing hydroxide ions (OH^-) at an electrode to form water.

Amperometric titration has the advantage over other types of titration due to its selectivity offered by the electrode potential, as well as by the choice of titrant. For instance, lead ions are reduced at a potential of –0.60 V (relative to the saturated calomel electrode), while zinc ions are not; this allows the determination of lead in the presence of zinc. A common environmental application of amperometric titration is the measurement of chlorine in drinking water.

17) **Use a table to compare the similarities and differences between potentiometry, coulometry, and voltammetry. Consider the following comparisons: (a) The electrical measurement (e.g., current, potential, and charge), (b) the types of cells (galvanic and electrolytic), (c) the fundamental equation employed for quantitative measurement (Nernst, Faraday, and Ohm's), (d) the ability for quantitative determination.**

Titrations	Type of measurement	Fundamental equation	Type of cell	Ability for quantitative determination
Potentiometry	Potential	Nernst	Galvanic	Yes
Coulometry	Charge	Faradays	Electrolytic	Yes
Voltammetry	Current	Ohm's law	Electrolytic	Yes

18) **Two major advantages of anodic stripping voltammetry (ASV) for metal analysis are (a) low detection limits (μg/L to ng/L), and (b) ability to analyze several elements in a sample run. Explain why.**

The low detection limit is due to preconcentration of metal from the solution into or onto a microelectrode (with a large surface area) by electrodeposition process. Subsequent to the preconcentration step, each metal is "stripped" off by applying increased potential (e.g., linear potential scan) so each metal is measured sequentially during a run.

19) **Define: (a) boundary potential, (b) limiting current, (c) equivalent points of titration, (d) alkaline error, (e) deposition potential.**

a) *Boundary potential* is the voltage that exists between an electrode and the solution in which it is immersed. This potential is referred to as standard electrode, such as the hydrogen electrode.

b) *Limiting current* is the limiting value of a Faradaic current that is approached as the rate the potential applied to an electrode is gradually increased. It is usually evaluated by subtracting the appropriate residual current from the measured total current.

c) *Equivalent points of titration* are the points where the number of moles of titrant (e.g., base) is equal to the number of moles of the analyte (e.g., acid). In potentiometric titration, for example, this is where the inflection point of the plot of potential (E) vs volume of titrant (V).

d) *Alkaline error* is a systematic error that occurs when glass electrodes are used to read the pH of an extremely alkaline solution. The electrode responds to sodium ions as though they were hydrogen ions, giving a pH reading that is consistently too low.

e) *Deposition potential* is the smallest potential that should be applied to an electrolytic cell to produce electrolytic deposition on an electrode.

Problems

1) **Calculate the cell potential for the cell described in Question 8: Sn | Sn^{2+} (0.06 M) ‖ Cl^- (0.001 M)| Cl_2 (1 atm) | Pt.**

Anode: $E^0 = -0.14$ V

Cathode: $E^0 = +1.36$ V

$$Q = \frac{[Sn^{2+}]}{[Cl^-]^2} = \frac{[0.06]}{[0.001]^2} = 60,000$$

$$E = E^0 - \frac{2.303RT}{nF} \log Q = 1.22 - \frac{0.0591}{2} \log 60,000 = 1.08 \text{ V}$$

2) **A typical automobile battery is rated as "100 ampere-hours," meaning the delivery of 1.0 A current for 100 hours. How many grams of Pb (atomic weight = 202.2) are oxidized for the full rating?**

$$m = \frac{Q}{F} \frac{M}{n} = \frac{1.0 \ A \times 100 \ hr \times 3600s/hr}{96,485 \ coulombs/mol \ e^-} \times \frac{207.2 \ g \ Pb/mol \ Pb}{2 \ mol \ e^-/mol \ Pb}$$

$$= 386.5 \text{ g Pb}$$

3) **The potential (E_1) in a 0.0015 mol/L F^- standard solution was measured to be 0.150 V by a fluoride electrode in reference to a saturated calomel electrode. For the same electrode, a voltage of 0.250 (E_2) was measured in a sample containing unknown concentration of F^-. Calculate the concentration of F^- in the unknown sample. (*Hint*: Use Eq. 12.9.)**
Applying Eq. 12.9, we can cancel out the constant and obtain the following:

$$E_1 - E_2 = \frac{2.303RT}{F}\left(pF_2 - pF_1\right)$$

By rearranging the above equation, we can get:

$$pF_2 = \frac{E_1 - E_2}{\dfrac{2.303RT}{F}} + pF_1 = \frac{0.150 \text{ volts} - 0.250 \text{ volts}}{\dfrac{2.303 \times \dfrac{8.314 \text{ J}}{\text{K} \times \text{mol}} \times 298 \text{ K}}{96,485 \dfrac{\text{J}}{\text{volts} \times \text{mol}}}} + (-\log 0.0015) = 1.133$$

$$-\log F_2 = 1.133$$

$$F_2 = 0.0736 \text{ mol/L}$$

Note that, in the above equations, F_1 and F_2 are the fluoride concentrations in the standard solution and unknown sample, respectively. F is the Faraday constant, which has a value of 96,485 coulombs $\times$ mol^{-1} or 96,485 J/(volts $\times$ mol). A temperature of 25°C (298 K) was assumed in the calculation. The lowercase "p" is the negative of the base 10 logarithm.

4) **Calculate the mass of Cu in gram that can be deposited from a solution of CuSO$_4$ when 0.1 Faraday electricity is used (atomic weight of Cu = 64, S=32, O=16).**

Apply Faraday's law:

$$m = \frac{Q}{F} \frac{M}{n}$$

where F, M, and n are constants representing, respectively, Faraday's constant (96,485 coulombs per mole electron), molecular weight of the analyte (g per mole analyte), and the number of mole of electrons per mole of analyte.

Since 1 Faraday electricity = 96,485 coulombs. The amount of electricity of 0.1 Faraday is 9648.5 coulombs. With this electricity, the amount of copper (Cu) in the solution can be determined:

$$m = \frac{9648.5 \text{ coulombs}}{96,485 \text{ coulombs/mol e}^-} \times \frac{63.5 \text{ g Cu/mol Cu}}{2 \text{ mol e}^-/\text{mol Cu}} = 3.175 \text{ g Cu}$$

Chapter 13

Questions

1) **Can the atomic weight listed in the periodic tables be used for calculating accurate molecular mass? Why or why not?**

No, because the atomic weight of an element, as commonly shown in the periodical table, is the weighted average of the atomic masses of its different isotopes. However, the exact mass of an isotopic species is calculated by summing the masses of the individual isotopes (rather than the weighted average of various isotopes) of the molecule.

2) **Which ion sources are operated at the atmospheric pressure without vacuum conditions? Why vacuum is important in mass analyzers?**

These ion sources are ESI, APCI, and APPI. Vacuum should be maintained in the mass analyzer so that the integrity of analyte ions remains unchanged without collision or any reactions.

3) **Give two examples of high-resolution mass spectrometer (HRMS).**

TOF, orbitrap

4) **Explain why 70 eV ionization energy is considered as the hard ionization.**

In electron ionization (EI), the ion source is a beam of highly energetic electrons with ionization energy of approximately 70 eV. This is considerably higher than the ionization energy of 5–15 eV (bond dissociation energies are even smaller) for most organic compounds. As a result of this excess energy, extensive fragmentations of molecules are taking place, hence "hard ionization" compared to other so-called soft-ionization sources.

5) **Which one of the following ion sources is likely the best in ionizing a nonpolar polyaromatic hydrocarbon: APCI, APPI, or ESI?**

APPI (see Figure 13.6)

6) **Give examples of applications for MALDI-TOF mass spectrometry.**

MALDI-TOF is especially suitable for large thermally labile biological molecules (nucleic acid, peptides, proteins), synthetic polymers, drugs, and metabolites. Another unique application is the direct identification of microbes.

7) **What are the uses of isotope ratio mass spectrometry and the isotope dilution mass spectrometry?**

The isotope dilution method is one specific type of internal standard calibration and quantitation method that takes advantage of the mass spectrometry through which the isotope ratio of an element or its corresponding compound is measured. In this method, an unknown sample is spiked with an enriched isotope (usually of minor abundance) of the element of interest.

Fundamentals of Environmental Sampling and Analysis, Second Edition. Chunlong Zhang.
© 2024 John Wiley & Sons, Inc. Published 2024 by John Wiley & Sons, Inc.
Companion Website: www.wiley.com/go/EnvironmentalSamplingandAnalysis2e

Isotope dilution mass spectrometry provides high accuracy and precision because the spiked isotope serves as both a calibration standard and an internal standard.

With the isotope ratio method, mass spectrometry measures specific isotopes of an element, and then the ratio of two or more isotopes of this element can readily be determined. By comparing the isotope ratio change, the isotope ratio information can be used in geological dating of rocks, forensic analysis, determining the sources of a contaminant (source appointment), and tropic transfer of pollutants in biological studies.

8) **Describe the difference between ion trap and orbitrap. What makes these ion sources high in resolution and accuracy in mass measurement compared to the quadrupole?**

The ion trap, or the commonly referred 3D ion trap, basically works on the same principle as a quadrupole mass analyzer, using static DC current and RF oscillating electric fields, where the parallel rods are replaced with two hyperbolic metal electrodes (end caps) facing each other, and a ring electrode placed halfway between the end cap electrodes; ions are trapped in a circular flight path based on the applied electric field.

Orbitrap is one of the configurations of the ion trap, but its hardware is configured differently. Orbitrap consists of an outer barrel-like electrode and a coaxial inner spindle-like electrode that traps ions in an orbital motion around the spindle. The image current from the trapped ions is detected and converted to a mass spectrum using the Fourier transform of the frequency signal.

Ion traps require a voltage ramp to detect different m/z's sequentially, while orbitraps detect all m/z's simultaneously.

9) **In time-of-flight tube, what will be the flight time change if the mass of the singly charged ion doubles?**

Time is proportional to the square root of mass (m), so the flight time will be 1.4 times longer when m is doubled.

10) **The magnetic sector mass analyzer is excellent in its resolution, accurate mass measurement, and reproducibility, but what makes it less commonly used compared to other mass analyzers?**

Bulky, expensive, and slow scan to match the GC or LC elution.

11) **What are the major similarities and differences between the mass spectrometers used in atomic mass spectrometry and those used in molecular mass spectroscopy?**

Atomic mass spectrometry (ICP-MS) is similar to molecular mass spectrometry (GC-MS and LC-MS) in that it consists of an ion source, a mass analyzer, and a detector. The identities (atoms or molecules) are determined by their mass-to-charge ratio (via the mass analyzer) and their concentrations are determined by the number of ions detected. Although mass analyzers and detectors are similar in principles between the two, it is the ion source in atomic mass spectrometry that differs the most from molecular mass spectrometry. In ICP-MS, the ion sources must atomize samples, or an atomization step must take place before ionization. The atomization and ionization of atoms in ICP-MS require very high temperature (7000 K in ICP torch) compared to low temperature (GC-MS) to room temperature (some LC-MS) to ionize molecules.

12) **Describe the ionization mechanisms for (a) electron ionization (EI), (b) chemical ionization (CI), (c) electrospray ionization (ESI), and (d) atmospheric-pressure chemical ionization (APCI).**

a) Electron ionization (EI): In the EI mode, a beam of high-energy electrons (approximately 70 eV) is produced by boiling electrons off a narrow strip or coil of wire made of tungsten–rhenium alloy. Because of the high energy compared to the bond dissociation energies (5–15 eV) of most organic molecules, this hard ionization usually produces many small pieces of fragment ions of the sample molecules, thereby offering rich structural information.

b) Chemical ionization (CI): CI uses a stream of gas such as CH_4, NH_3, isobutene in both GC-MS and LC-MS. Since lower energy is used to bombard the analyte molecule than the electron ionization, fragmentation is minimized in this soft ionization mode. Mass spectrum is usually simpler with few peaks.

c) Electrospray ionization (ESI): As shown in Fig. 13.4, analyte solution is sprayed (nebulized) into a chamber at atmospheric pressure in the presence of a strong electrostatic field and heated drying gas. The resulting charged spray of fine droplets then passes through a desolvating capillary, where evaporation of the solvent and attachment of charge to the analyte solvent take place. As the droplets become smaller as a consequence of evaporation of the solvent, their charge density becomes greater and desorption of ions into the ambient gas occurs.

d) Atmospheric-pressure chemical ionization (APCI): The APCI method is analogous to CI (commonly used in GC-MS) where primary ions are produced by corona discharge on a solvent spray. In APCI (Fig. 13.5), the LC eluent is sprayed through a heated (typically 250–400°C) vaporizer at atmospheric pressure. The heat vaporizes the liquid, and the resulting gas-phase solvent molecules are ionized by electrons discharged from a corona needle. The solvent ions then transfer charge to analyte molecules through chemical reactions (chemical ionization). The analyte ions pass through a capillary sampling orifice into the mass analyzer.

13) **Why do the molecular ions of certain compounds not show up under EI but show up under CI and ESI?**
This is because EI is the hard ionization compared to the soft ionization with CI and ESI. Under the impact of high-energy electrons in EI, most molecules cannot remain intact (molecular ions) but undergo fragmentation.

14) **Illustrate the principles of quadrupole-based mass analyzer. Why is it also commonly referred to as a mass filter?**
A quadrupole system consists of four metal rods approximately 20 cm in length and 1 cm in diameter. The quadrupole separates ions by setting up the correct combination of voltages and radio frequencies to guide ions of selected m/z between the four rods. Ions do not match the selected m/z pass out through spaces between the rods and are ejected from the quadrupole. With this regard, the quadrupole is also referred to as a mass filter.

15) **Describe the operational principles of quadrupole-based mass analyzer.**
The quadrupole consists of four metal rods approximately 20 cm in length and 1 cm in diameter. Opposing rods are connected in pairs to both a direct current (DC) generator and a radio frequency (RF) generator (Fig. 13.14). The quadrupole separates ions by setting up the correct combination of DC voltages (U) and radio frequencies having an oscillating RF voltage (V) to guide ions of selected m/z between the four rods.

The two generators in quadrupole are operated in such a way that m/z becomes linearly dependent on only one variable, either U or V. Thus, the quadrupole mass analyzer can be easily controlled, and scan can be quickly operated in the range of 0 up to about 800 Da. The ions that do not match the selected m/z pass through spaces between the rods are ejected from the quadrupole and removed through the vacuum pump.

16) **Briefly explain the working principles of the following mass spectrometry: QqQ, Q-TOF, APCI-ITTOF, LC-ESI-QTOF, FAB-B, GC-FI-TOF.**

QqQ: Refers to triple quadrupole mass spectrometry (TQMS or QqQ). QqQ uses three quadrupoles in series. In the initial quadrupole (Q1), the mixture of analyte ions is separated and certain ions (precursor ions) are allowed to pass to Q2. This first quadrupole is used to select user-specified analyte ions, usually the molecular-related (i.e., $(M+H)^+$ or $(M-H)^-$) ions. In the collision cell (Q2), collision-induced dissociation (CID) takes place. Within the collision cell the precursor ions, also known as parent ions, are then bombarded with an inert gas (Xe, Ar, etc.) and are further fragmented into product ions of different charges and masses. These product ions, also known as daughter ions, are then run through an additional quadrupole (Q3) to further separate the ions for subsequent detection of the ion fragments.

Q-TOF: Refers to quadrupole-time-of-flight (Q-TOF). Q-TOF uses a quadrupole (four parallel rods), a collision cell, and a time-of-flight unit to produce mass spectra. It is similar to that of a triple-quadrupole mass spectrometer (QqQ), although the third quadrupole has been replaced by a time-of-flight tube.

APCI-IT-TOF: Refers to atmospheric pressure chemical ionization-ion trap-time of flight mass spectrometry. Ions are generated by atmospheric pressure chemical ionization (APCI) sources. These continuous sources of ions are focused from the source into the ion trap (IT) by a combination of electrostatic lenses and a split RF octopole ion guide. The ions are then analyzed by time-of-flight (TOF) mass analyzer.

LC-ESI-QTOF: Refers to liquid chromatography-electrospray ionization-quadrupole-time-of-flight. The eluent from HPLC is pumped through a stainless-steel capillary needle, which is then sprayed (nebulized) into a chamber at atmospheric pressure in the presence of a strong electrostatic field and heated drying gas. The resulting charged spray of fine droplets then passes through a desolvating capillary, where evaporation of the solvent and attachment of charge to the analyte solvent take place. As the droplets become smaller as a consequence of evaporation of the solvent, their charge density becomes greater and desorption of ions into the ambient gas occurs. These ions are then passed through a capillary sampling orifice into a quadrupole (four parallel rods), a collision cell, and then ions are detected through time-of-flight unit to produce high-resolution mass spectra.

FAB-B: Refers to fast atom bombardment – magnetic sector mass spectrometry. In FAB-B, fast atoms are obtained by passing accelerated rare gas (argon or xenon) atoms from an ion source or gun. The beam of energetic fast atoms is used to bombard and ionize analytes, which are typically solubilized in a viscous liquid (e.g., glycerol, thioglycerol). The ions generated from FAB are typically detected using a double-focusing mass spectrometer, in this case here; it is magnetic sector mass spectrometry. The double-focusing mass spectrometer requires an electrostatic analyzer as well as a magnetic sector analyzer to focus ions with various directions and velocities so that these ions can be separated according to the *m/z* ratios. As such, FAB-B is less attractive than ESI and MALDI mass spectrometry.

GC-FI-TOF: Refers to gas chromatography-field ionization – time-of-flight mass spectrometry. Here, analytes are separated in GC, and vapor of the analytes from GC column eluent is ionized (with little fragmentation) on a tungsten wire emitter through which a high voltage is applied. The ions are detected by time-of-flight mass analyzer based on the flight times.

17) **Explain the terms related to mass spectrometry: multiple reaction monitoring (MRM), collision-induced dissociation (CID), and MS^4.**

Multiple reaction monitoring (MRM): Also called selected reaction monitoring (SRM). It is a method used in tandem mass spectrometry in which an ion of a particular mass is selected in the first stage of a tandem mass spectrometer and an ion product of a fragmentation reaction of the precursor ions is selected in the second mass spectrometer stage for detection.

Collision-induced dissociation (CID): CID is a cell in a mass spectrometry where fragmentation of selected ions is induced in the gas phase. The selected ions (typically molecular ions or protonated molecules) are usually accelerated by applying an electrical potential to increase the ion kinetic energy and then allowed to collide with neutral molecules (often helium, nitrogen or argon). During collision, some of the kinetic energy is converted into internal energy, which results in bond breakage and the fragmentation of the molecular ions into smaller fragments.

MS^4: MS^n refers to sequential mass spectrometry. The initial steps are the same as in QqQ for the formation of product ions. The difference is that the product ions from QqQ are not consumed by the detector, but are further trapped and allow another isolation and fragmentation to be performed resulting in the MS^3 spectrum. This process can be repeated a few times resulting in a series of MS^n spectra where n represents the number of isolation-fragmentation-measurement cycles. MS^n requires ion-trap mass analyzer to allow the refragmentation of product ions.

18) **Explain what are the differences between "SCAN" and "SIM" in GC-MS. Why is "SIM" needed to achieve high sensitivity and lower detection limits?**

All mass filters can be operated in two different modes. In the selected ion monitoring (SIM) mode, a selected ion(s) of the target analyte at a specific m/z value(s) are collected and detected subsequently. The SIM mode provides a plot of abundance of selected ions vs retention time. In the scan mode, all ions from the analyte (in a range of m/z values) are collected and analyzed. SIM is often used for quantitative analysis to achieve a high sensitivity, because more ions can be collected in a given time period. The scan mode detects all isotopes and fragment ions and contains all structural information needed for identification purposes. (Note: Some recent GC-MS models can do both SCAN and SIM in the same run.)

19) **Explain why interface is important and essential for all hyphenated mass spectrometers. Describe the interface used in ICP-MS, GC-MS (EI), and LC-MS (electrospray).**

An interface is a very critical component of all hyphenated mass spectrometers. Each hyphenated mass spectrometer, however, is unique in the interface coupling of two instruments and the method of ionization. In ICP-MS, the ICP torch, used to atomize and ionize materials, requires a high operational temperature (~6000 K), whereas the mass spectrometer is operated at room temperature and requires the vacuum condition to avoid collisions with any gas molecules before ions can reach the detector. This task is accomplished through an interface between ICP and MS, and the use of vacuum pumps to remove nearly all gas molecules in the space between the interface and the detector.

The interface between GC and MS is relatively simple. GC equipment can be directly interfaced with rapid-scan mass spectrometers with a proper flow adjustment. For capillary columns, the flow rate is usually small enough to feed directly into the ionization chamber of the mass spectrometer. If packed columns are used, a jet separator can remove the carrier gas.

The interface between HPLC and MS is much more challenging and is the key to success. HPLC systems use high pressure for needed separation efficiency and a high load of liquid flow and hence high gas load. (For example, a common flow of liquid at 1 mL/min, when converted to the gas phase, is 1 L/min.) The MS part, on the other hand, requires high vacuum and elevated temperate, and does not tolerate high flow of introduced sample. Two interface technologies became available in the 1990s to address such problems in coupling LC with MS (detailed in textbook).

20) **Compare the similarities and the differences between ICP-OES and ICP-MS with regard to the operational principles and the instrumental components.**

In ICP-OES, elements are ionized by ICP and then measured based on the emission of elemental ions by an optical device at a wavelength characteristic of the element in the UV-VIS range. In ICP-MS, elements are ionized by ICP in the same way as in ICP-OES. However, rather than separating emission light according to their wavelengths, the mass spectrometer separates ions according to their mass-to-charge ratios (m/z). Because ions are counted in ICP-MS, the numbers (abundance) of ions are used as the basis for quantitative measurement. In ICP-MS, the mass spectrum is therefore a plot of abundance vs m/z, rather than the optical UV-VIS spectrum in ICP-OES, which is a plot of light emission vs wavelength.

ICP-OES and ICP-MS have the same sample introduction system and torch to generate ions from their respective elements to be analyzed. In detecting ions with ICP-MS, ions from the ICP torch are first focused with an electrical field. This is analogous to the optical lens used in ICP-OES to bend the light beams of various wavelengths. The mass filter (quadrupole) in ICP-MS is equivalent to a monochromator in ICP-OES used to separate out the wavelengths of the light emitted by elemental ions.

ICP-MS has a much higher price range than ICP-OES; however, it presents several analytical benefits compared to ICP-OES and other common elemental analysis instrumentation. These include: (a) better sensitivity than graphite furnace AAS; (b) rapid multielement quantitative analysis; (c) wider dynamic range; (d) lower detection limit, and (e) much simpler spectrum as compared to ICP-OES (mass spectrum vs optical emission spectrum). The optical emission spectrum may have hundreds of emission lines for most heavy elements, demanding a high-resolution optical device to minimize potential spectral interference. In ICP-MS, however, there are only 1–10 natural isotopes for all elements in the mass spectrum.

21) **What are the functions of (a) lens, (b) quadrupole, and (c) dynode detector in ICP-MS?**

   a) Lens: The main function is to *focus* ions generated from ICP torch. An ion beam is focused before it enters the mass analyzer. It comprises one or more ion lens components, which electrostatically steer the analyte ions from the interface region into the mass separation device.

   b) Quadrupole: Sometimes known as the ion optics, its main function is to *separate* ions according to various m/z ratios. A quadrupole system consists of four metal rods approximately 20 cm in length and 1 cm in diameter. The quadrupole separates ions by setting up the correct combination of voltages and radio frequencies to guide ions of selected m/z between the four rods. Ions do not match the selected m/z pass out through spaces between the rods and are ejected from the quadrupole.

   c) Dynode: The main function is to *detect* (count) ions. The ions exiting from the mass spectrometer strike the active surface of the detector (known as a dynode) and generate a measurable electronic signal. The active surface of the detector releases an electron each time an ion strikes it. The ion exiting from the quadrupole strikes the first dynode that releases electrons and starts the amplification process. The electrons released from the first dynode strike a second dynode where more electrons are released. This cascading of electrons continues until a measurable pulse is created. By counting the pulses generated by the detector, the system counts the ions that hit the first dynode.

22) **Explain molecular ions, fragment ions, isotopic clusters, and base peak in a molecular mass spectrum.**

A molecular ion peak is the unfragmented (intact) molecule, which is often the largest (heaviest) peak among the highest m/z group (except for any isotopic peaks). The molecular ion represents the parent molecule minus one electron. Since most of the ions formed in a mass spectrometer have a single charge, the m/z value is equivalent to mass itself. Therefore, molecular ions have an m/z corresponding to the molecular weight.

A fragment ion is an electrically charged dissociation product of an ionic fragmentation. Such an ion may dissociate further to form other electrically charged molecular or atomic moieties of successively lower formula weight.

An isotopic cluster is a group of fragmented ions having the same molecular formula but different formula weight due to the differing composition of the isotopes made of the molecule. For example, a cluster of the molecular ions for 1,4-dichlorobenzene ($C_6H_4Cl_2$) is made of ions having m/z of 146, 148, and 150 for $C_6H_4{}^{35}Cl_2$, $C_6H_4{}^{35}Cl^{37}Cl$, $C_6H_4{}^{37}Cl_2$, respectively.

A base peak is defined as the largest peak in the mass spectrum. Its relative abundance is arbitrarily defined as 100%. The most intense ion is assigned an abundance of 100, and it is referred to as the base peak. The base peak is the tallest peak because it represents the most common fragment ion to be formed – either because there are several ways in which it could be produced during fragmentation of the parent ion, or because it is a particularly stable ion.

23) **Which one(s) of the following compounds should not have the isotope peaks in the mass spectra besides the peaks related to H and C isotopes: (a) fluoroethane, (b) 1,1,1-trichloroethane, (c) tribromomethane, (d) ethyl iodide.**

(a) and (d), because both (b) and (c) contain Cl or Br with their isotopes and thus have isotopic clusters in mass spectra.

24) **What is the abundance ratio of ^{35}Cl to ^{37}Cl and ^{79}Br to ^{81}Br? How do these isotopic ratios affect the mass spectra of compounds containing Cl and Br?**

The natural abundance of ^{35}Cl and ^{37}Cl are 75.53% and 24.47%, respectively. The natural abundance of ^{79}Br and ^{81}Br are 50.69% and 49.31%, respectively. In other words, the two chlorine atoms have an approximate isotopic ratio of $^{35}Cl/^{37}Cl = 3:1$, and the two bromine atoms have an approximate isotopic ratio of $^{79}Br/^{81}Br = 1:1$. Such information can be critical to the interpretation of mass spectra particularly when atoms such as chlorine and bromine are present in the organic compounds.

25) **Explain how $\delta^{18}O$ in the ice samples taken near the poles can be used to tell the climate change.**

The cold regions are depleted in the heavy ^{18}O but remain high in ^{16}O because the temperature affects evaporation/condensation of water cycle throughout the globe.

26) **A mass spectrum of a halogen compound shows peaks at m/z (relative abundance) of 188 (12%), 186 (9%), 109 (72%), 107 (75%). What halogen atom is likely in this compound?**

$188-107 = 81$, $188-109 = 79$, and $186-107 = 79$. This matches the atomic weight of two bromine isotopes: ^{81}Br and ^{79}Br. Also, the m/z is 2 units apart with almost equal relative abundance, which confirms the presence of Br. It is actually 1,1-dibromoethane, with molecular weight of 190 ($C_2Br^{81}H_4$), 188 ($C_2Br^{81}Br^{79}H_4$), and 186 ($C_2Br^{79}H_4$)

27) **Which of the following species will not respond to the mass analyzer and will be removed by the vacuum prior to reaching the detector: (a) $CH_3CH_2CH_2$, (b) $^+CH_2CH_3$, (c) $(CH_3CH_2CH_3)^+$, (d) $[H_2PO_4]^-$?**

(a) is the only neutral species and will be removed by vacuum.

28) **The mass spectrum of pentane from an electron ionization is shown below:**

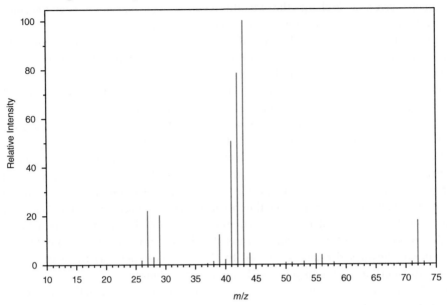

a) **Indicate which peak is the molecular ion peak. Which one is the base peak?**

b) **What causes the line at *m/z* of 57, *m/z* of 43, and *m/z* of 29?**

c) **Can you propose a fragmentation mechanism based on the above mass spectrum?**

d) **Write all the reactions involved in the fragmentation of pentane.**

a) Pentane has a chemical formula of C_5H_{12} (i.e., $CH_3(CH_2)_3CH_3$) and molecular weight of 72. The peak of molecular ion, $CH_3(CH_2)_3CH_3{}^+$ or $CH_3(CH_2)_3CH_3{}^{+\cdot}$, has an *m/z* of 72 as can be seen from the mass spectrum. (Here the dot in $CH_3(CH_2)_3CH_3{}^{+\cdot}$ represents a single unpaired electron. That is one half of what was originally a pair of electrons – the other half is the electron that was removed in the ionization process.) The base peak (*m/z* = 43) was due to $CH_3CH_2CH_2{}^{+\cdot}$. This base peak has a relative intensity of 100, and the peak height of all other ions is measured relative to the base peak.

b) The ions corresponding to *m/z* of 57, 43, and 29, are $CH_3(CH_2)_3{}^{+\cdot}$, $CH_3CH_2CH_2{}^{+\cdot}$, and $CH_3CH_2{}^{+\cdot}$, respectively.

c) The mass spectrum suggests that pentane can lose a terminal CH_3 group, which results in a methyl radical (neutral) and an ion with *m/z* of 57 (72–15 = 57). The bond between C_2 and C_3 can also be dissociated, which results in an ethyl radical (neutral) and an ion with *m/z* of 43 (72–15–14 = 43). Also, pentane molecules can be dissociated between C_3 and C_4, which generates a propyl radical (neutral) and an ion with *m/z* of 29 (72–15–14–14 = 29). The other lines in the mass spectrum are more difficult to explain. For example, lines with *m/z* values 1 or 2 less than one of the above lines are often due to loss of one or more hydrogen atoms during the fragmentation process.

d) The major reactions involved in the fragmentation of pentane are as follows:

$$CH_3CH_2CH_2CH_2CH_3{}^+ \rightarrow CH_3CH_2CH_2CH_2{}^+ (m/z = 57) + CH_3\cdot(\text{methyl radical})$$

$$CH_3CH_2CH_2CH_2CH_3{}^+ \rightarrow CH_3CH_2CH_2{}^+ (m/z = 43) + CH_2CH_3\cdot(\text{ethyl radical})$$

$$CH_3CH_2CH_2CH_2CH_3{}^+ \rightarrow CH_3CH_2{}^+ (m/z = 29) + CH_2CH_2CH_3\cdot(\text{propyl radical})$$

29) **The mass spectrum of 2-chloropropane (C_3H_7Cl, MW = 78) below shows four clusters with the major peaks at 78, 63, 43 (base peak). Explain the following:**

a) **The cluster around m/z of 78 shows m/z peaks of 78, 79, 80, and 81.**

b) **The possible fragment(s) of the base peak at 43 and 27.**

c) **The peak at m/z of 63, 64, 65, and 66.**

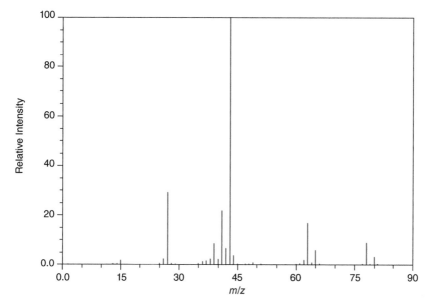

Chlorine (Cl) has two isotopes: 76% Cl^{35} and 24% Cl^{37}. Thus, M = $C_3H_7Cl^{35}$ = 78; M + 2 = $C_3H_7Cl^{37}$ = 80; M+1 with one ^{13}C = 81.

a) 78 (M), 80 (M+2), 79 (M with one ^{13}C), 81 (M+1 with one ^{13}C).

b) 43 = M − 35 or M+2 − 37; 27 = M − 35 − 15 or M+2 − 37 − 15.

c) 63 = M−15 (CH_3); 65 = M+2 − 15; 64 = M−15 with one ^{13}C; 66 = M+1 − 15 with one ^{13}C.

30) **The mass spectra of (1) ethanol, (2) tribromomethane, and (3) trichloromethane are given below. Match the compounds with the mass spectra.**

(a)

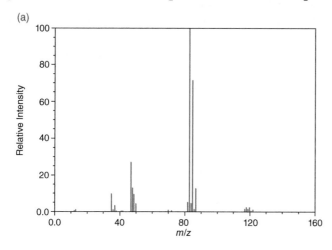

(b)

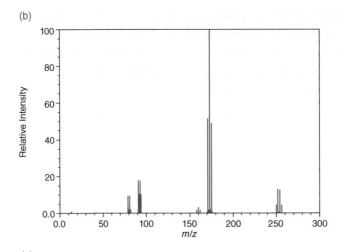

(c)

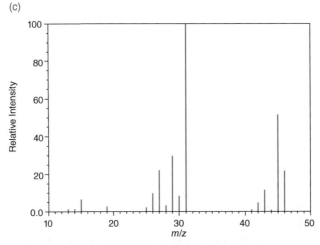

The match of mass spectra to these three distinctly different compounds should be straightforward: (a) trichloromethane, (b) tribromomethane, (c) ethanol. In short, this is because ethanol should not have any isotopic cluster, whereas the approximately equal abundance of isotopic peaks should be assigned to a Br-containing compound and the other with different relative abundances should be assigned to a Cl-containing compound.

More specifically on the mass spectra (a) trichloromethane, $C^{35}Cl_3H = 118$, $C^{37}Br_3H = 124$; hence the first cluster will have m/z of 118, 120, 122, and 124 (2 mass apart). The second cluster is the result of the loss of one Cl, with m/z in the range of $118-35 = 83$ to $118-37 = 87$. The same can be said for the third cluster for a m/z in the range of $83-35 = 48$ to $87-37 = 50$. Also note the distinctly different relative abundances of ions in the same clusters, because the natural abundance of Cl is 76% ^{35}Cl and 24% ^{37}Cl.

For mass spectrum (b) tribromomethane: $C^{81}Br_3H = 256$, $C^{79}Br_3H = 250$; hence, the first cluster will have m/z of 250, 252, 254, and 256 (2 mass apart). The second cluster is the result of the loss of one Br, with m/z in the range of $250-79 = 171$ to $250-81 = 175$. The same can be said for the third cluster for an m/z in the range of $171-79 = 92$ to $175-81 = 94$. Different from the mass spectrum in (a), the relative abundances of ions in the same clusters are approximately the same because the natural abundance of Br is 51% ^{79}Br and 49% ^{81}Br.

For spectrum (c), ethanol has a molecular weight of 46, which can lose one CH_3 to become m/z of 31 (the base peak) or lose the hydroxyl (OH) to become m/z of 29.

31) **The mass spectra of tetrachloroethylene under (a) GC-EI-MS and (b) GC-CI-MS are given in the figure below (Reprinted with permission from Richer et al. (2006). Copyright (2006) American Chemical Society). Methane was used as the ionization gas in the CI mode. Propose fragmentation/reaction pathways.**

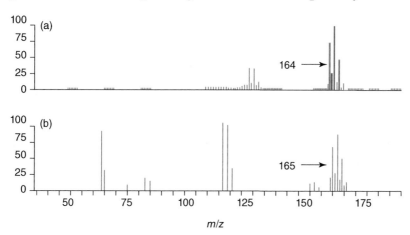

The molecular mass of $C_2{}^{35}Cl_4H_2 = 166$, and $C_2{}^{37}Cl_4H_2 = 174$. Hence, ions with m/z of 166, 168, 170, 172, and 174 are likely fragments, although m/z of 166 corresponding to $C_2{}^{35}Cl_4H_2$ would be the most likely. The GC-EI-MS spectrum with a base peak at 166 reflects these isotopic peaks.

In the chemical ionization (CI) mode, fragmentation such as $(M+1)^+$ and $(M-1)^+$ ions are more likely observed. These ions correspond to m/z of 167 and 165, respectively.

As shown in the EI mass spectrum, EI ionization produces the molecular ion at $m/z = 166$, and the fragmentation of this molecular ion further produces a fragment ion at m/z of 91 as follows:

$$C_2{}^{35}Cl_4H_2 + e^- \rightarrow C_2{}^{35}Cl_4H_2{}^+ (166) + 2e^-$$

$$C_2{}^{35}Cl_4H_2{}^+ \rightarrow C_2{}^{35}Cl_4{}^+ (164) + 2H$$

The isotopic peak with m/z of 166 is a base peak, and the peak at m/z of 164 is the peak with the second highest abundance.

In comparison, chemical ionization generally produces $(M+1)^+$ and $(M-1)^+$ ions. The $(M+1)^+$ ion at m/z of 167 is produced from the **proton transfer**:

$$CH_5{}^+ + C_2{}^{35}Cl_4H_2 (166) \rightarrow CH_4 + C_2{}^{35}Cl_4H_3{}^+ (167)$$

A $(M-1)^+$ peak at m/z of 165 is produced by the following **hydride transfer** reaction:

$$C_2H_5{}^+ + C_2{}^{35}Cl_4H_2 (166) \rightarrow C_2H_6 + C_2{}^{35}Cl_4H^+ (165)$$

$CH_5{}^+$ and $C_2H_5{}^+$ in the above two reactions represent two primary ions produced from methane as the gaseous reagent in the ion source.

32) **Is it true that the peak with the highest m/z in a mass spectrum is always the M or M+1 ion?**

Not true, because ^{13}C or ^2H will also be likely the ion with the highest m/z.

33) **Can you tell whether the following pollutants have an even or odd nominal molecular mass without doing the calculation: (a) pentachlorophenol (C_6HCl_5O), (b) nitrobenzene ($C_6H_5NO_2$), (c) 2,4-dintrotoluene ($C_7H_6N_2O_4$), (d) 2,4-dimethylphenol ($C_8H_{10}O$)?**

According to nitrogen rule, only (b) has the odd number of N, hence the odd molecular weight.

34) **Give a list of the major US EPA methods for the GC-MS analysis of (a) VOCs and (b) SVOCs in drinking water, wastewater, and wastes.**

The US EPA's standard GC-MS methods for the analysis of VOCs can be found in Methods 524 and 624 for drinking water and wastewater, respectively. The equivalent GC-MS methods for SVOCs can be found in Methods 525 and 625 for drinking water and wastewater, respectively. The GC-MS-based waste methods in SW-846 include 8260B (VOCs), 8270C (SVOCs), and a particular high-resolution GC-MS method for the analysis of dioxin compounds.

35) **In Chapter 6, Table 6.7 lists some reported methods, mostly mass spectrometry, for the analysis of several groups of emerging contaminants. Use the same format (i.e., chromatography-ion source-mass analyzer), conduct a literature search (e.g., using Google Scholar, Web of Science, etc.), and tabulate the reported mass spectrometric methods for (a) pesticides and their transformation products, (b) polybrominated diphenyl ethers (PBDEs) as ingredients in the brominated fire retardants.**

Students' answers may vary. A good review paper can be used for this purpose. For example, Richardson SD (2012) wrote a good review paper on "Environmental mass spectrometry: emerging contaminants and current issues" in *Anal Chem* 84:747–778.

a) LC/MS/MS, UPLC/MS/MS, Q-TOF-MS
b) GC-NCI-MS, GC/HRMS, GC-IT-MS, LC/APPI/MS/MS, LC/API-MS/MS

36) **Conduct a literature search for the reported use of MALDI-TOF-MS in enabling the identification of water-borne pathogens (bacteria and virus) in drinking water that was not possible with the traditional microscopic techniques. Explain how in principle can MALDI-TOF-MS fingerprint microorganisms at the genera and oftentimes at the species level.**

MALDI-TOF-MS can detect a large number of proteins simultaneously. Because the patterns are reproducible, it became possible to compare unknown samples with a database via data analysis algorithms that enable the automatic detection, indexing and statistical comparisons of the matching patterns of different proteins in various microorganisms.

37) **Propose a specific mass spectrometric method appropriate for the following analytical purpose: (a) differentiate whether sediment Pb is due to anthropogenic source (such as a smelting plant) or natural background; (b) improve the recovery and detection accuracy for the analysis of pesticides in water; (c) identify bacteria directly without the use of optical microscopy; (d) verify if a nearby manufacturing facility contributes to a BTEX plume in an area of groundwater pollution.**

(a) ICP-IRMS, (b) GC-IDMS, (c) MALDI-TOF, (d) GC-IRMS

38) **Search for literature from peer-reviewed journal papers on nontargeted analysis of the environmental pollution of your interested topic and list the types of mass spectrometry including the type of chromatography, ion source, mass analyzer, as well as sample preparation and concentration, if any.**

Students' answers may vary. Conducting a keyword search such as "nontargeted analysis environment" in Google Scholar or a specific journal such as *Environmental Science and Technology* will output many research papers in peer-reviewed journals.

Problems

1) **Which molecular formula has the exact molecular mass of 78.0468: (a) C_3H_7Cl, (b) C_6H_6, (c) $C_2H_3FO_2$, (d) $(CH_3)NCl$.**

Using accurate mass in Table 13.1, C = 12.0000, H = 1.0078; Cl = 34.9689, F = 18.9984, O = 15.9949, N = 14.0031. (a) 78.0235; (b) 78.0468; (c) 78.0116, (d) 79.0188. The answer is (b).

2) **Iron has stable isotopes ^{54}Fe, ^{56}Fe, ^{57}Fe, and ^{58}Fe with isotopic mass (abundances) of 53.9396 (5.845%), 55.9349 (91.754%), 56.9354 (2.119%), and 57.9333 (0.286%), respectively. What is its nominal mass and atomic weight (conventional), and the exact atomic mass of iron?**

The nominal mass is taken from the most abundant isotope in its whole number; thus, it is 56.

The atomic weight is the abundance weighted average of all the stable isotopes:

Atomic weight = 53.9396×0.05845+55.9349×0.91754+56.9354×0.02119+57.9333×0.00286 = 55.8474

In the periodic table, the atomic weight is shown as 55.845(2). The difference is due to the round-off error of the abundance.

3) **Calculate the nominal mass and the exact mass of glucose $C_6H_{12}O_6$.**

The nominal mass for M = $(6 \times 12) + (12 \times 1) + (6 \times 16) = 180$, whereas the exact mass is $(6 \times 12.0000) + (12 \times 1.0078) + (6 \times 15.9949) = 180.0630$.

4) **Calculate the resolution needed for a mass spectrometer to separate two ions with exact masses of 268.1704 and 268.1902. Would it need a high-resolution mass spectrometry?**

$(m_1 + m_2)/2 = (268.1704 + 268.1902)/2 = 268.1803$

$\Delta m = (268.1902 - 268.1704 = 0.0198$

$R = 268.1803/0.0198 = 13544$

5) **Calculate the molar energy (kJ/mol) for the electrons if the accelerating potential is maintained at 20 eV.**

$E = e\,V\,N = (1.60\times10^{-19}\,C/e^-)(20\text{ V})(6.02\times10^{23}\,e^-/\text{mol}) = 1.93\times10^6\,J/\text{mol}$
$= 1{,}930\,kJ/\text{mol}$

6) **A TOF-MS is operated at 4,800 volts, and the flight tube length is 1.2 m. (a) Calculate the flight time in μs needed for an ion of 270.0807 Da to reach the detector. (b) Estimate the time difference in ns for the TOF to distinguish between two singly charged ions with Δm of 0.0180.**

a) By applying Eq. 13.3, we have:

$$t = d\sqrt{\frac{m}{2zeV}} = 1.2\text{ m}\sqrt{\frac{270.0807\,\text{Da}\times1.66\times10^{-27}\,\text{kg}/\text{Da}}{2\times1\times1.6\times10^{-19}\,C\times4800\text{ V}}} = 2.05\times10^{-5}\,s = 20.50\,\mu s$$

b) For singly charged ions, $\Delta m = 0.0180$, $\Delta m/z = 0.0180$. Substituting this m/z into Eq. 13.3, we have:

$$t = d\sqrt{\frac{m}{2zeV}} = 1.2\text{ m}\sqrt{\frac{0.0180\,\text{Da}\times1.66\times10^{-27}\,\text{kg}/\text{Da}}{2\times1\times1.6\times10^{-19}\,C\times4800\text{ V}}} = 1.67\times10^{-7}\,s = 0.167\,\mu s$$
$$= 167\,ns$$

7) **A magnetic sector mass analyzer with a fixed radius of curvature of 0.12 m is operated at an accelerating voltage of 2,580 V while varying the strength of magnetic field. To scan the m/z in a range of 50–500 for singly charged ions, what range of magnetic strength is required?**

We rearrange Eq. 13.7,

$$B = \frac{1}{r}\sqrt{\frac{2mV}{ze}}$$

Using consistent SI unit, for $m/z = 50$:

$$B = \frac{1}{r}\sqrt{\frac{2mV}{ze}} = \frac{1}{0.12\,\text{m}}\sqrt{\frac{2\times50\,\text{Da}\times1.66\times10^{-27}\,\frac{\text{kg}}{\text{Da}}\times2580\,\text{V}}{1\times1.60\times10^{-19}\,\text{C}}} = 0.431\,\text{T}$$

For $m/z = 500$:

$$B = \frac{1}{r}\sqrt{\frac{2mV}{ze}} = \frac{1}{0.12\,\text{m}}\sqrt{\frac{2\times500\,\text{Da}\times1.66\times10^{-27}\,\frac{\text{kg}}{\text{Da}}\times2580\,\text{V}}{1\times1.60\times10^{-19}\,\text{C}}} = 1.363\,\text{T}$$

8) **Estimate the number of carbon atoms if the ratio of [M+1]/[M] is 12.1.**

The number of C atoms $= 12.1/1.1 = 11$ C atoms.

9) **An unknown compound has a molecular ion peak with a relative abundance of 55.92% and M+1 peak with a relative abundance of 4.31%. Estimate the number of C atoms in this compound.**

The number of C atoms $= 4.31/(55.92\times0.011) = 7.007 \approx 7$

10) **A trace amount of vanadium (V) in solid sample can be measured directly by laser-ablation ICP coupled with isotope dilution mass spectrometry (ICMS). A 5 g solid sample containing an unknown V concentration was thoroughly mixed with a 0.5 g spike containing 10 μg/g V enriched with ^{50}V (64.35% ^{51}V and 35.65%% ^{50}V). Vanadium has two naturally abundant isotopes: ^{51}V at 99.76% and ^{50}V at 0.24%. The isotope ratio of the homogeneously mixed solid sample and the spike was measured to be at 0.95. Determine the concentration of V (^{50}V + ^{51}V) in this solid sample.**

$C_s = 10$ μg/g; $m_s = 0.5$ g, $m_x = 5$ g, $A_s = 0.6435$, $B_s = 0.3565$, $A_x = 0.0024$, $B_x = 0.9976$, $R_m = 0.95$.

Applying Eq. 13.9, we obtain:

$$C_x = C_s \frac{V_s}{V_x} \frac{A_s - R_m B_s}{R_m B_s - A_x} = 10\,\frac{\mu g}{g}\,\frac{0.5\,\text{g}}{5\,\text{g}}\,\frac{0.6435 - 0.95\times0.3565}{0.95\times0.9976 - 0.0024} = 0.3225\,\frac{\mu g}{g}$$

11) **An Antarctic snow sample has an ^{18}O/^{16}O ratio of 0.0018732 and a ^{2}H/^{1}H ratio of 0.00010873. What are its δ^{18}O and δ^2H values relative to the Vienna Standard Mean Ocean Water? Report in per mil and explain whether the snow sample is isotopically light or heavy.**

Vienna Standard Mean Ocean Water (VSMOW: ^{18}O/^{16}O = 0.0020052, ^{2}H/^{1}H = 0.00015576)

$$\delta^{18}\text{O} = (0.0018732/0.0020052 - 1) = -0.065829 = -65.830$$

$$\delta^2\text{H} = (0.00010873/0.00015576 - 1) = -0.301939 = -301.940$$

Both are negative, meaning that the Antarctic water is isotopically light, owing to the fact that light water is evaporated from the warmer area and then deposited in the cold Antarctic region.

Chapter 14

Questions

1) **Explain why is NMR resolution increased with the increasing strength of the magnetic field (B_0). Why is ^{1}H-NMR more sensitive than ^{13}C-NMR when an equal number of nuclei is compared.**

This is because the resonant frequency (ν) is directly proportional to the strength of the magnetic field (Eq. 14.2). As a result, the spectral resolution of an NMR will be increased when the magnetic field strength (B_0) is increased. The modern day 600 MHz or even 900 MHz NMR is in significant contrast with regard to sensitivity and resolution compared to the 30 MHz NMR in the early days.

The NMR responsive isotope ^{13}C has only a 1.1% abundance and its gyromagnetic ratio (γ) is also much smaller than that of ^{1}H (6.7283×10^7 rad/T/s vs 2.6752×10^8 rad/T/s). This implies that the ^{13}C-NMR is much less sensitive than ^{1}H-NMR (relative sensitivity: 1.59×10^{-2} vs 1).

2) **Explain why a nucleus with more shielding resonates at a lower radio frequency, and why such a nucleus is located in the upfield.**

From the upfield to downfield, the resonance frequency is in an increasing order. Whether a nucleus appears in an upfield or a downfield depends on the degree of the shielding, which in turn depends on the electronic environment of the nucleus. In general, we can state that:

Protons in *electron-rich* molecular environment are *more shielded*. This "shielding" means an electronegative functionality (such as F, Cl, Br) pulls the electron toward it. These shielded protons sense a *smaller magnetic field* and thus come into resonance at a *lower frequency*. These ^{1}H will appear on the right-hand side of the NMR spectrum (upfield).

Conversely, protons in an *electron-poor* molecular environment are *less shielded*. These less shielded or deshielded protons sense a *larger magnetic field* and thus come into resonance at a *higher frequency*. These ^{1}H will appear on the left-hand side of the NMR spectrum (downfield).

3) **Explain the difference between diamagnetic shielding and spin–spin coupling.**

Protons are surrounded by an electronic cloud of charge due to adjacent bonds and atoms. In an applied magnetic field (B_0), electrons circulate and produce an induced field (B_i), which opposes the applied field. The effective field at the nucleus will be $B = B_0 - B_i$. The nucleus of the proton is said to be experiencing a *diamagnetic shielding*, because it "senses" a lower magnetic field.

Fundamentals of Environmental Sampling and Analysis, Second Edition. Chunlong Zhang.
© 2024 John Wiley & Sons, Inc. Published 2024 by John Wiley & Sons, Inc.
Companion Website: www.wiley.com/go/EnvironmentalSamplingandAnalysis2e

The *spin–spin coupling* (or spin–spin splitting) occurs when a proton that we are looking at (H_A) is near another nonequivalent proton (H_B). In half of the molecules, the H_A proton will be adjacent to an H_B aligned with the field and in the other half the H_A proton will be adjacent to an H_B aligned against the field. Thus, half the H_As in the sample will feel a slightly larger magnetic field than they would in the absence of H_B and half will feel a slightly smaller magnetic field. Thus, we will observe two absorptions for the H_A proton. (The same is true for H_B.) This splitting of the H_A resonance into two peaks is termed "spin–spin coupling" or "spin–spin splitting" and the distance between the two peaks (in Hz) is called the "coupling constant" (usually represented by the symbol J). The spin–spin coupling is transmitted through the electrons in the bonds and so depends on the bonding relationship between the two hydrogens.

4) **Can the signals of ^{1}H and ^{13}C be measured at the same time with NMR spectrometers?**

The resonance frequencies of ^{13}C nuclei are lower than those of protons in the same applied field. For example, in an instrument with a 7.05 Tesla magnet, protons resonate at about 300 MHz, while carbons resonate at about 75 MHz. This is fortunate, as it allows us to look at ^{13}C signals using a completely separate "window" of radio frequencies. ^{1}H and ^{13}C are not measured at the same time.

5) **Identify elements that are NMR-active or NMR-inactive: ^{12}C, ^{15}N, ^{16}O, ^{23}Na, ^{31}P.**

^{15}N, ^{23}Na, and ^{31}P are NMR-active.

6) **Explain how does magnetic field strength affect differently on chemical shift and coupling constant.**

The absolute resonance frequency depends on the applied magnetic field (Eq. 14.2), but the chemical shift is *commonly* considered to be independent of external magnetic field strength. (Note: There is a report that states otherwise. Refer to: Ella Wren, 2020, Recent Evidence that NMR chemical shifts depend on magnetic field strength, *Chemistry World*.)

The spacing in the frequency unit (Hz), also called the coupling constant (J), determines the extent of coupling between two nuclei. Since coupling is caused solely by the internal molecular forces, the magnitudes of J are not dependent on the operating frequency of the NMR spectrometers. This characteristic coupling constants (J) can then be used to infer the number and type of bonds that connect the coupled protons as well as the geometric relationship of protons.

7) **Determine how many signals would you expect from the following compounds in the ^{1}H-NMR spectrum: (a) tetrachloroethylene, (b) trichloroethene, (c) 1,1-dichloroethene, (d) 1,2-dichloroethene, (e) vinyl chloride (refer to Appendix C for their chemical structures).**

The numbers of ^{1}H-NMR signals of the chlorinated hydrocarbons (CHCs) compounds are summarized in the table below. The letters in the chemical structure in an alphabetically ascending order denote an increasing chemical shift in the downfield. Note that tetrachloroethylene has no ^{1}H-NMR signals because it lacks any protons. Also note that the isomeric 1,2-dichloroethene and 1,2 dichlorethene have the same number of chemically equivalent protons, but the ^{1}H-NMR spectrum is very different. For 1,2-dichloroethene, the ^{1}H-NMR spectrum clearly shows the characteristic coupling constant that represents the geometric relationship of the two neighboring protons.

Compound and formula	Structure	Number of chemically equivalent carbons (NMR signals)
Tetrachloroethylene (C_2Cl_4)		0
Trichloroethene (C_2HCl_3)		1
1,1-Dichloroethene ($C_2H_2Cl_2$)		1
1,2-Ddichloroethene ($C_2H_2Cl_2$)		1
Vinyl chloride (C_2H_3Cl)		3

8) **Determine how many signals would you expect from the following compounds in the ^{13}C-NMR spectrum: (a) benzene, (b) toluene, (c) ethylbenzene, (d) *o*-xylene, (e) *m*-xylene, (f) *p*-xylene (refer to Chapter 2 for their chemical structures).**

The numbers of ^{13}C-NMR signals of the BTEX compounds are summarized in the table below. The numbers in the chemical structure in an increasing order denote an increasing chemical shift in the downfield.

Compound and formula	Structure	Number of chemically equivalent carbons (NMR signals)
Benzene (C_6H_6)		1
Toluene ($C_6H_5CH_3$)		5
Ethylbenzene ($C_6H_5CH_2CH_3$)		6
o-Xylene ($C_6H_4CH_3CH_3$)		4
m-Xylene ($C_6H_4CH_3CH_3$)		5
p-Xylene ($C_6H_4CH_3CH_3$)		3

9) **For $CH_3CH_2CH_2Cl$, which proton signal would you expect to be at the lowest frequency (upfield) in ^1H-NMR? For the same compound, which carbon signal would you expect to be at the lowest frequency (upfield) in ^{13}C-NMR?**

The protons in the terminal methyl group (CH_3) will have the lowest frequency (upfield) in the ^1H-NMR, and the protons immediately adjacent to Cl will have the highest chemical shift (downfield). The three protons in the methyl group are the most shielded because methyl group is the furthest from an electron withdrawing functional group Cl. As a result, the nuclei of these three protons "sense" a lower magnetic field, which in turn requires a lower resonance frequency (hence the lower chemical shift in the downfield).

In ^{13}C-NMR, the carbon in the terminal methyl group (CH_3) will also have the lowest frequency (upfield) in the ^{13}C-NMR. This is also because of the effect of electron-withdrawing Cl furthest to the C in the methyl group.

For $CH_3CH_2CH_2Cl$, both ^1H-NMR and ^{13}C-NMR will give three groups of NMR signals on the NMR spectrum. However, one big difference is the split signals in the ^1H-NMR but no split signals in the ^{13}C-NMR.

10) **What would be the order of the chemical shift (δ values) for the compounds: CH_3Br, CH_3Cl, CH_3F, CH_3I in ^1H-NMR? Why?**

All of these four compounds will have only one signal (singlet), because each compound has only one type of chemically equivalent proton. Since the electronegativity is in an increasing order of I < Br < Cl < F, the proton in CH_3F is the least shielded (fluorine atom will draw electrons away from the methyl group). The ^1H nucleus in CH_3F will sense a higher magnetic field, which will in turn require a higher frequency of the rf radiation to bring into the resonance of the ^1H. The order of the chemical shift (δ values) will be $CH_3F > CH_3Cl > CH_3Br > CH_3I$. From AIST database, the chemical shifts corresponding to CH_3F, CH_3Cl, CH_3Br, and CH_3I are 4.10 ppm, 3.052 ppm, 2.682 ppm, and 2.165 ppm, respectively.

11) **Which one of the following will exhibit its ^1H-NMR signal at the highest chemical shift: CHF_3, CH_2F_2, CH_3F?**

Because the electronegative F atom(s) pull H away from the nuclei, the proton(s) in these three molecules are in the increasing order of shielding: $CHF_3 < CH_2F_2 < CH_3F$. The chemical shift will be in a decreasing order of $CHF_3 > CH_2F_2 > CH_3F$. Checking with https://sdbs.db.aist.go.jp, the chemical shift are 6.25 ppm, 5.46 ppm, and 4.10 ppm for CHF_3, CH_2F_2, and CH_3F, respectively.

12) **Why tetramethylsilane (TMS) is typically used as a reference standard for NMR spectroscopy?**

The ^1H-NMR of tetramethylsilane (TMS), $Si(CH_3)_4$, appears at the lowest frequency (upfield) than most signals; it therefore can be used as an NMR standard. The reason for this lowest frequency is that silicon (Si) is the least electronegative in organic molecules. Consequently, the methyl protons of TMS are in more electron-dense environment than most protons in organic molecules. Thus, the protons in TMS are the most shielded, and their nuclei sense the lowest magnetic field.

13) **What are the respective uses of two deuterated solvents ($CDCl_3$ and D_2O) in NMR?**

Deuterated chloroform ($CDCl_3$) is used for organic compounds, and deuterated water (D_2O) is used for highly polar and ionic compounds in NMR.

14) **Which group of protons in $CH_3CH_2CH_2CH_2NO_2$ will have the highest chemical shift?**

Two protons immediately adjacent to NO_2 will have the highest chemical shift (downfield). These two protons are the least shielded because NO_2 is an electron-withdrawing functional group. As a result, the nuclei of these two protons "sense" a larger magnetic field, which in turn requires a higher resonance frequency (hence the highest chemical shift in the downfield).

15) **In a high-resolution ^{1}H-NMR of a compound $CH_3CH_2CH_2Cl$, how many split signals would you expect from the methyl group, the protons in the $-CH_2Cl$ group, and the protons of the central $-CH_2-$ group?**

$CH_3CH_2CH_2Cl$ has three types of H, hence three groups of NMR signals in ^{1}H-NMR spectrum. The methyl group is the furthest from the electron-withdrawing Cl, and therefore it is the least shielded (downfield), followed by the middle H and the H adjacent to Cl (upfield).

The first group of signal (downfield) due to protons in the methyl group will have $2+1 = 3$ split signals (triplet), because it has 2 H immediately adjacent to the proton in methyl group. The NMR signal due to the central $-CH_2-$ group will have $5+1 = 6$ split signals because it has a total of 5 H immediately adjacent to it. The third group of signal (upfield) due to protons in the $-CH_2Cl$ group will have $2+1 = 3$ split signals (triplet) because there are 2 H immediately adjacent to it.

16) **Identify the most shielded and least shielded protons in (a) 2-chlorobutane, (b) 1,2,2-tribromopropane.**

a) $CH_3CHClCH_2CH_3$: The electronegative Cl will withdraw electrons from protons, so the least shielded is the CH in C2 position, and the most shielded is the CH_3 on the C4 position that is further away from Cl. Checking with https://sdbs.db.aist.go.jp, the chemical shifts are 3.972 ppm for the least shielded CH in C2, and 1.024 ppm for the most shielded CH_3 on the C4 position.

b) $CH_2BrCBr_2CH_3$: The electronegative Br will withdraw electrons from protons, so the least shielded is the CH_2 in C1 position, and the most shielded is the CH_3 in C3 position. The SDBS database does not have ^{1}H-NMR spectrum for 1,2,2-tribromopropane, but a similar chlorinated compound 1,2,2-trichloropropane $CH_2ClCCl_2CH_3$ in the SDBS database reveals 4.04 ppm for CH_2 in C1 position and 2.23 ppm for CH_3 in C3 position.

17) **Assign the chemical shifts (a) δ 1.87 and 3.86 to the appropriate protons of 1,2-dibromo-2-methylpropane, (b) δ 0.96, 1.33, 1.53, 1.59, 3.64 ppm to the appropriate protons of 2-chloropentane.**

(a) 1,2-dibromo-2-methylpropane (b) 2-chloropentane

D(A)   3.860

D(B)   1.870

(a) 1,2-dibromo-2-methylpropane (b) 2-chloropentane

18) **Assign the chemical shifts (a) δ 21.4, 125.4, 128.3, 129.1, and 137.8 ppm to the appropriate carbons of toluene. (b) δ 26.5, 128.3, 128.6, 133.0, 137.2, and 197.9 ppm to the appropriate carbons of acetophenone.**

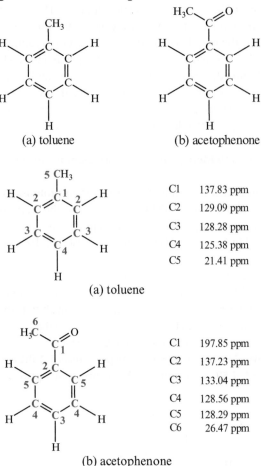

(a) toluene

C1	137.83 ppm
C2	129.09 ppm
C3	128.28 ppm
C4	125.38 ppm
C5	21.41 ppm

(a) toluene

C1	197.85 ppm
C2	137.23 ppm
C3	133.04 ppm
C4	128.56 ppm
C5	128.29 ppm
C6	26.47 ppm

(b) acetophenone

19) **Predict the appearance of the high-resolution proton ^{13}C-NMR spectrum of (a) ethyl ethanoate, (b) methacrylonitrile.**

(a) ethyl ethanoate

(b) methacrylonitrile

a) Ethyl ethanoate has four peaks corresponding to four types of carbon atoms. The most obvious peaks are the ester carbonyl at around 170 ppm and the deshielded –CH_2 attached to O at around 60 ppm. The carbon atoms of the two methyl groups away from the ester carbonyl show the ^{13}C-NMR signals at the upfield (20–30 ppm) as shown in the ^{13}C-NMR spectrum.

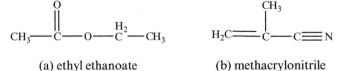

C1	171.08 ppm
C2	60.44 ppm
C3	21.00 ppm
C4	14.28 ppm

(a) ethyl ethanoate

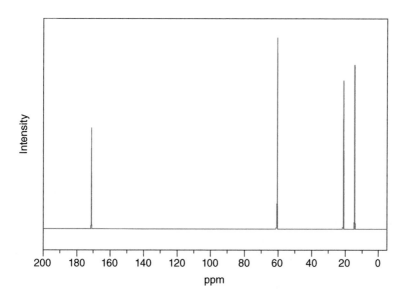

b) Methylacrylonitrile has four peaks corresponding to four types of carbon atoms. The methyl C is the most shielded with a chemical shift of around 20 ppm. The carbon in C≡N is around 120 ppm, and the in the alkene CH_2 double-bonded with a nearby C (C=C) is around 130 ppm.

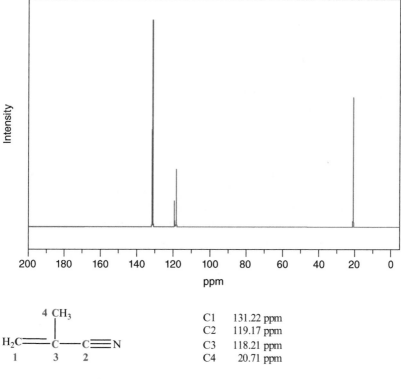

C1	131.22 ppm
C2	119.17 ppm
C3	118.21 ppm
C4	20.71 ppm

(b) methacrylonitrile

20) **Describe the advantages and disadvantages of ^{13}C-NMR as compared to ^{1}H-NMR.**

First, the signals in ^{13}C-NMR are not normally split by neighboring carbons and the interpretation of ^{13}C-NMR spectra is easier than ^{1}H-NMR spectra. The reason for the lack of splitting is that the probability of two ^{13}C carbon atoms being next to each other in a molecule is very small ($1.1\% \times 1.1\% = 0.0121\%$; here 1.1% is the natural abundance of ^{13}C). Second, ^{13}C-NMR is less sensitive because of the low abundance of ^{13}C and its gyromagnetic constant (Table 14.1). To alleviate this problem, more samples or a longer run time is needed. Third, unlike ^{1}H-NMR, the area under a ^{13}C-NMR spectrum is *not* proportional to the number of atoms giving rise to the signal, implying that integration cannot be routinely used for ^{13}C-NMR unless special techniques are used. Lastly, the range of the chemical shift in ^{13}C-NMR spectra is much wider than ^{1}H-NMR (220 ppm vs 12 ppm), so the potential signal overlap is minimized. This is an advantage of ^{13}C-NMR.

21) **Describe the difference in atomic structure and properties between ^{13}C (used in NMR) and ^{14}C (used in radio-labeled analysis).**

^{13}C has a nucleus containing 6 protons and 7 neutrons. It is present naturally at 1.1% natural abundance. It is not radioactive, but magnetically active with a spin quantum number of 1/2 (like ^{1}H) and therefore detectable by NMR.

^{14}C has a nucleus containing 6 protons and 8 neutrons. The natural abundance is only 1 part per trillion. It is radioactive, can have beta decay. Its half-life is 5,720 years. This long half-life is the basis for the radiocarbon dating method to date archaeological, geological, and hydrogeological samples.

22) **Explain why are AES and XPS complementary techniques for surface chemical analysis?**

The signal of Auger electron for a given element shifts if a different X-ray source is used. This is because the binding energy of Auger electron changes as the source of incident X-ray source is varied. In XPS, however, the binding energy of the ejected electrons remains the same regardless of the types of X-ray source. The binding energy of the ejected electrons in XPS is unique to the element and material properties to be studied. Therefore, using a different X-ray source, the spectral overlaps of an element can be resolved by employing two different X-ray sources (e.g., Mg source and Al source). This is achieved by commercially available single instrument for both AES and XPS measurement, which makes AES and XPS two complementary rather than competitive techniques.

23) **Explain why are XRF and Auger electron emission competing processes?**

As shown in Fig. 14.12(b) and (c), both XRF and AES involve the ejection of electrons upon the irradiation of the incident X-rays. In XRF, the vacancy of the ejected electron in the inner orbital is filled by another electron in the outer shell, and in the meantime, extra energy is emitted in the form of fluorescence. In AES, the vacancy of the ejected electron in the inner orbital is filled by another electron in the outer shell, but extra energy brings out the ejection of the Auger electron from the outer shell. These are two competing processes. The relative rates of fluorescence and Auger emission depend on the atomic number of the element. High atomic numbers favor XRF, whereas Auger emission occurs primarily for atoms of low atomic numbers.

24) **Schematically show an MNN electron transition of an Auger emission spectroscopy.**

For low atomic number elements, the most probable transitions occur when a K-level electron is ejected by the primary beam, an L-level electron drops into the vacancy, and another L-level electron is ejected. Higher atomic number elements have LMM and MNN transitions that are more probable than KLL.

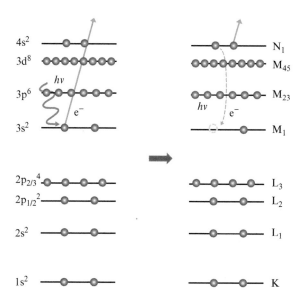

MNN Auger electron emission

25) **How can one tell whether a peak in an electron spectrum is due to the Auger electrons?**

The peak of Auger electron emission will shift if a different X-ray source is used (e.g., Mg source or Al source). This is different from peaks in XPS because the binding energy of the ejected electrons remains the same regardless of the types of X-ray source.

26) **By referencing Fig. 14.11, (a) describe the information that reveals respectively from an XPS survey spectra and the high-resolution spectra. (b) Explain why the peaks are shown in the order from the left to right as of Zn 2p, Co 2p, O 1s, N 1s, and C 1s. (c) Explain why SO_4^{2-} peak is shown in the left of the SO_3^{2-} peak.**

a) Survey XPS spectra reveal the atomic species, their atomic percentages and characteristic binding energies. High-resolution XPS (HR-XPS) spectra provide information on the chemical oxidation state and bonding of those elements.

b) An element with a higher atomic number will have a higher positive nuclear charge, thus the attraction between the electron to the nucleus is increased. The energy for the attraction between an electron and a nucleus is the binding energy. For the elements shown in Fig. 14.11(a), the binding energies of these electrons are in the increasing order of C 1s < N 1s < O 1s < Co 2p < Zn 2p, which is consistent with the increasing bounding energy (from right to left) on the Survey XPS spectra.

c) The magnitude of binding energy of valence electrons to the nucleus is affected by oxidation state of the element. Since an atom with lost electron(s) exhibits a positive charge, it will be more difficult to remove an electron from that atom, thus the binding energy will be higher as the oxidation state increases, e.g., $SO_4^{2-} > SO_3^{2-}$. Sulfur (S) has 6 valence electrons, $1s^2 2s^2 sp^6 3s^2 3p^4$. While S in SO_4^{2-} has an oxidation number of 6, i.e., S(VI), S in SO_3^{2-} has an oxidation number of 4, S(IV).

27) **Describe the difference (e.g., mechanism, uses, resolution, sample preparation) between (a) SEM and TEM, (b) STM and AFM.**

a) SEM and TEM: Both are electron microscopy, i.e., using a high voltage of electron beam to "see" objects just like "light" is used to see objects under light microscopy. SEM provides images of the external morphological and topographical information analogous to the human eye, whereas transmission electron microscopy (TEM) provides images of the internal structure of a solid sample. In SEM, an incident beam of electrons interacts with a solid sample and the electrons are bounced back as the **backscattering electrons** and **secondary electrons** as the secondary beam to be detected and converted into signals of an image. In comparison, electrons in TEM transmit (pass) through a specimen to form an image of the internal structure. The difference between these two electron microscopes can be better realized with an analogy of SEM as a flashlight in a dark room, whereas TEM as a light transmitted through an image on a film.

SEM has a resolution of 1–20 nm, whereas TEM has a resolution of 0.1 nm. For SEM, samples are first coated with a metal that readily reflects electrons. This coating is also essential to conducting surface for electrons to avoid charging of the sample. Electrons are emitted from a wire (usually tungsten or lanthanum hexaboride) that has been super-heated by an electric current. In TEM, in order for the electrons to pass through a sample in TEM, the sample should be sectioned to be very thin, generally less than 150 nm because electrons cannot readily penetrate sections much thicker than 200 nm. In high-resolution TEM imaging, the thickness should even be below 30 nm.

b) STM and AFM: Both are scanning probe microscopes (SPMs) used to determine the 2-D (STM) or 3-D (AFM) topography of a surface with an ultra-high resolution (atomic scale). Scanning tunneling microscope (STM) can resolve features on an atomic scale on the surface of a conducting solid sample. In STM, the surface of a sample is scanned by a very fine metallic tip. This sharp tip is moved up and down, based on the topographical structure of the surface, by monitoring the tunneling currents between the tip and sample surface in order to maintain the tip at a constant distance from the sample surface. A tunneling current is a current that passes through a medium (vacuum, a nonpolar liquid) that contains no electrons.

Atomic force microscopes (AFM) allow for the resolution of individual atoms on both conducting and insulating surfaces. This is advantageous over STM because STM requires a conducting surface to be examined. In AFM, a sharp tip mounted on a soft lever is scanned across the sample surface, while the tip is in contact with the surface. The atomic force acting on the tip changes according to the sample topography, resulting in a varying deflection of the lever. The deflection of the lever is measured by means of laser beam deflection of a microfabricated cantilever and subsequent detection with a double-segment photodiode.

Samples for AFM must be adhered rigidly and properly dispersed on a substrate (mica, silicon, glass, or metal disc), followed by activation and binding of the sample to the substrate, rinsing with deionized water and finally drying before visualization. AFM measurements can be made under ambient pressure or in liquid. Thus, underwater imaging for biological specimens can be obtained without distortion.

28) **What are the major pros and cons of SEM, TEM, STM, and AFM for the use of surface analysis?**

The table below is a brief but not exhaustive summary of major pros and cons.

	Pros	Cons
SEM	• $10–10^6$ times magnified images • Large area for surface details and composition	• Cannot image wet samples • Limited use to image nonconductive samples

TEM	• High resolution down to individual atoms • Analysis of quality, shape, size, and density	• Large and quite expensive • Require thin samples and electron transparent for electrons to pass through • Sample must be able to withstand vacuum
STM	• High resolution (0.1 nm lateral; 0.01 nm depth) • Atomic level atom surface arrangement • 3-D surface profile • Versatile on air, liquid, and solid	• Require very clean and stable (vibration free) surface • Difficult to use, and specific technique required • Expensive
AFM	• Provide 3D surface profile • No special sample treatment (coatings) • Can work well in ambient air or liquid. • Atomic resolution in ultra-high vacuum	• Limited single scan image size • Slow rate of scanning can lead to thermal drift • Cannot normally measure steep walls or overhangs

29) **Identify the X in the following radioactive decay:**

a) $^{14}_{6}C \rightarrow ^{14}_{7}N + X$

b) $^{40}_{19}K + X \rightarrow ^{40}_{18}Ar + X - ray$

c) $^{238}_{92}U + ^{234}_{90}Th + X$

These radioactive decays are:

a) $X = \beta^-$ (i.e., $^{0}_{-1}\beta$): $^{14}_{6}C \rightarrow ^{14}_{7}N + ^{0}_{-1}\beta$ (beta radiation)

b) $X = \beta^-$ (i.e., $^{0}_{-1}\beta$): $^{40}_{19}K + ^{0}_{-1}\beta \rightarrow ^{40}_{18}Ar + X-ray$ (electron capture and X−ray radiation)

c) $X = ^{4}_{2}He$: $^{238}_{92}U \rightarrow ^{234}_{90}Th + ^{4}_{2}He$ (alpha radiation)

30) **Compare ^1H, ^2H, and ^3H in terms of the number of protons, neutrons, mass number, and stability.**

These three naturally occurring isotopes of hydrogen are compared in the table below. The symbols D and T (instead of ^2H and ^3H) are sometimes used for deuterium and tritium:

	^1H (proton)	^2H (deuterium)	^3H (tritium)
Number of protons	1	1	1
Number of neutrons	0	1	2
Mass number	1	2	3
Stability	Stable	Stable	Unstable (radioactive)

31) **What types of shields are needed to protect against the exposure of: (a) α-particles, (b) β-particles, and (c) γ-rays.**

The penetration power is in an increasing order of α, β, and γ due to their varying degrees of mass and energy levels. The protection is warranted with the shielding of a piece of paper only for α particles or a thick layer of aluminum foil for β particles, but it requires several cm thick lead brick to fully protect from γ-ray exposure.

32) **For low-energy β emitters such as ^{14}C, ^{3}H, and ^{35}S, which one of the radioactivity measuring devices is desirable?**

Liquid scintillation counting is the method of choice for measuring β emitters, particularly weak β emitters such as ^{14}C, ^{3}H, and ^{35}S.

33) **What are the primary differences between γ-ray and X-ray?**

X-rays are emitted from processes outside the nucleus, but gamma rays originate inside the nucleus. X-rays are also generally lower in energy and therefore less penetrating than gamma rays.

34) **Provide a list of available methods for the analysis of radon gas in air.**

The following tests are briefly mentioned in the textbook:

- Use Do-It-Yourself Kits, and have samples sent to commercial labs for analysis.
- Battery-operated continuous radon monitors (more expensive than passive samplers).
- Passive samplers:
  - charcoal canisters (short-term: 2–7 days) and then measured using liquid scintillation.
  - alpha-track detectors (long-term: 4 months to a year).
  - charcoal liquid scintillation devices.
  - electret ion chamber detectors.

35) **What immunoassay methods have been available and adopted in the SW-846 methods? Give a list of such compounds that can be tested for field screening purpose.**

The immunoassays recommended by the US EPA include several specific contaminants, including pentachlorophenol (4010A), 2,4-dichlorophenoxyacetic acid (4015A), polychlorinated biphenyls (PCBs) (4020), petroleum hydrocarbons (4030), PAHs (4035), toxaphene (4040), chlordane (4041), DDT (4042), explosives (4050), RDX (4051), triazine herbicides as atrazine (4670). Details can be found in the 4000 method series of the SW-846 methods.

36) **Justify which of the following compound: (a) benzoic acid, (b) benzyl acetate, or (c) 2,6-dichlorotoluene matches the ^{13}C-NMR spectrum shown below.**

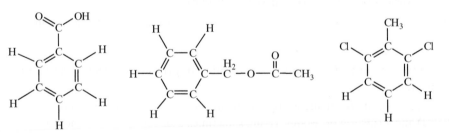

(a) benzoic acid   (b) benzyl acetate   (c) 2,6-dichlorotoluene

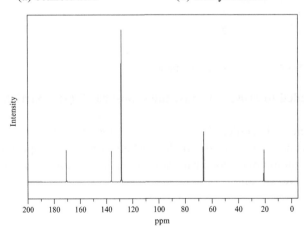

Benzyl acetate is the match.

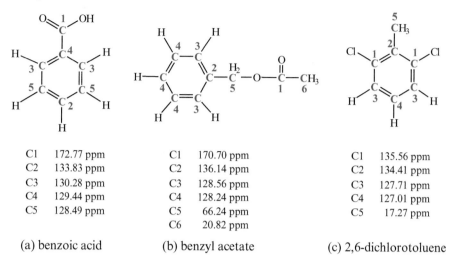

C1	172.77 ppm		C1	170.70 ppm		C1	135.56 ppm
C2	133.83 ppm		C2	136.14 ppm		C2	134.41 ppm
C3	130.28 ppm		C3	128.56 ppm		C3	127.71 ppm
C4	129.44 ppm		C4	128.24 ppm		C4	127.01 ppm
C5	128.49 ppm		C5	66.24 ppm		C5	17.27 ppm
			C6	20.82 ppm			

(a) benzoic acid   (b) benzyl acetate   (c) 2,6-dichlorotoluene

37) **Match the following three proton NMR spectra with *p*-xylene, ethyl alcohol, and ethyl benzene.**

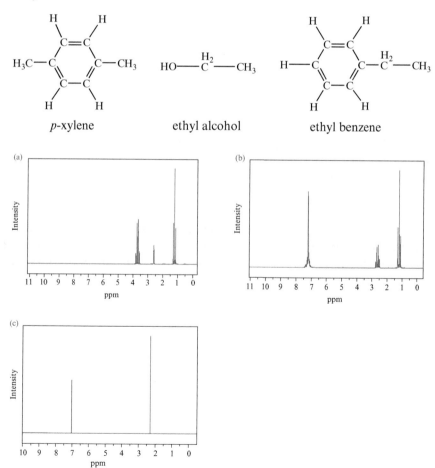

p-xylene   ethyl alcohol   ethyl benzene

a) Ethyl alcohol (CH_3CH_2OH): It has three types of equivalent protons. The least shielded CH_3 (1.2 ppm) should have 3 splits, CH_2 (3.7 ppm) should have 5 splits, and OH (2.61 ppm) should have 3 splits.

b) Ethyl benzene: The most shielded CH_3 (1.22 ppm) should have 3 splits, and the CH_2 (2.63 ppm) next to it should have several splits (in high-resolution NMR), the 5 aromatic protons (ArH) is the least shielded (7.0–7.45 ppm).

c) p-Xylene (Ar-$(CH_3)_2$): It has only two types of equivalent protons, leading to only two peaks in the ^{1}H-NMR as shown in (c): the ArCH$_3$ is around 7 ppm and ArH around 2.3 ppm.

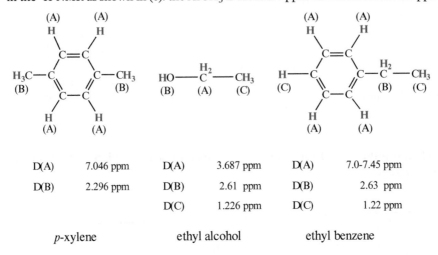

D(A)	7.046 ppm		D(A)	3.687 ppm		D(A)	7.0-7.45 ppm
D(B)	2.296 ppm		D(B)	2.61 ppm		D(B)	2.63 ppm
			D(C)	1.226 ppm		D(C)	1.22 ppm

p-xylene   ethyl alcohol   ethyl benzene

Problems

1) **If an ^{1}H-NMR spectrometer is equipped with a magnet with a magnetic field (B_0) of 7.05 T, calculate the required operating frequency of the spectrometer in megahertz (MHz). Use the tabulated value in Table 14.1 for the gyromagnetic ratio (γ).**

$$v = \frac{\gamma}{2\pi} B_0 = \frac{2.6752 \times 10^8 \text{ rad/T/s}}{2\pi} \times 7.05 \text{ T} = 300{,}321{,}019 \text{ rad/s} = 300 \text{ MHz}$$

2) **Refer to the above question, what will be the required frequency in MHz if this is an ^{13}C-NMR spectrometer? If the operating frequency is fixed at 200, 300, 500, 600 MHz, what will be the required magnetic field strengths (B_0) at each of these frequencies?**
For ^{13}C, the gyromagnetic ratio (γ) is 6.7283 ×10^7 rad/T/s. Hence, the required frequency is:

$$v = \frac{\gamma}{2\pi} B_0 = \frac{6.7283 \times 10^7 \text{ rad/T/s}}{2\pi} \times 7.05 \text{ T} = 75{,}532{,}667 \text{ rad/s} = 75.5 \text{ MHz}$$

If other operating frequencies are desired, then the required magnetic strength (B_0):

$$B_0 = \frac{2\pi v}{\gamma} = \frac{2\pi \times 200{,}000{,}000 \frac{\text{rad}}{\text{s}}}{6.7283 \times 10^7 \frac{\text{rad}}{\text{T} \times \text{s}}} = 18.67 \text{ tesla (for } v = 200 \text{ MHz)}$$

$$B = \frac{2\pi v}{\gamma} = \frac{2\pi \times 300{,}000{,}000 \frac{\text{rad}}{\text{s}}}{6.7283 \times 10^7 \frac{\text{rad}}{\text{T} \times \text{s}}} = 28.00 \text{ tesla (for } v = 300 \text{ MHz)}$$

$$B_0 = \frac{2\pi\nu}{\gamma} = \frac{2\pi \times 500,000,000\,\frac{\text{rad}}{\text{s}}}{6.7283\times10^7\,\frac{\text{rad}}{\text{T}\times\text{s}}} = 46.67 \text{ tesla (for } \nu = 500 \text{ MHz)}$$

$$B_0 = \frac{2\pi\nu}{\gamma} = \frac{2\pi \times 600,000,000\,\frac{\text{rad}}{\text{s}}}{6.7283\times10^7\,\frac{\text{rad}}{\text{T}\times\text{s}}} = 56.00 \text{ tesla (for } \nu = 600 \text{ MHz)}$$

3) **Calculate the chemical shift (δ) in ppm if a proton signal is 1150 Hz higher than that of the TMS standard. The ^{1}H-NMR spectrometer was run at a frequency of 600 MHz.**
Chemical shift (δ) measures the difference between the resonance frequency of the nucleus and a reference standard relative to the operating frequency of an NMR spectrometer. This quantity is calculated by:

$$\delta = \frac{\nu_{\text{signal}} - \nu_{\text{reference}}}{\nu_{\text{spectrometer}}} \times 10^6 = \frac{1150 \text{ Hz} - 0 \text{ Hz}}{600 \text{ MHz} \times \frac{10^6 \text{ Hz}}{1 \text{ MHz}}} \times 10^6 = 1.917 \text{ ppm}$$

4) **In a 300-MHz ^{1}H-NMR spectrometer, bromomethane CH_3-Br (10% wt in CCl_4) registers a signal at 804.60 Hz. (a) What is the chemical shift of the protons? (b) What would be the resonance frequency of the NMR signal if the spectrometer is operated at 500 MHz? (b) Rank the order of ^{1}H-NMR chemical shifts of the following compounds: CH_3-Br, CH_3-F, CH_3-N, and $(CH_3)_3$-Si.**

a) Apply Eq. 14.3, we have:

$$\delta = \frac{\nu_{\text{signal}} - \nu_{\text{reference}}}{\nu_{\text{spectrometer}}} \times 10^6 = \frac{804.60 \text{ Hz} - 0 \text{ Hz}}{300 \times 10^6 \text{ Hz}} \times 10^6 = 2.682 \text{ ppm}$$

b) According to Eq. 14.2, the frequency (ν) is proportional to the magnetic field strength (B_0). Hence, the frequency will be 804.60×500 Hz/300 Hz = 1341 Hz.

c) The electronegativity is in an increasing order of Br < N < F. The most electronegative F will withdraw electrons from the nucleus the most so that the nucleus is most deshielded (less shielded) and will feel the strongest frequency to resonate. Hence, proton signal for CH_3-F will show on the far left of the downfield or the higher chemical shift. Chemical shift in increasing order is $(CH_3)_4$–Si < CH_3–Br < CH_3–N < CH_3–F.

5) **An ejected electron from Co $2p$ has a kinetic energy of 694 eV using Al $K\alpha$ ($\lambda = 0.83393$ nm) as the source for X-ray in XPS. The electron spectrometer has a work function of 12.5 eV. (a) Calculate the energy of the incident X-ray source in eV. (b) Calculate the kinetic energy of the ejected electron in XPS. (c) If Mg $K\alpha$ is used, what would be the binding energy of Zn $2p$? (d) If this elected electron is an Auger electron using the irradiated source of Al $K\alpha$, what would its kinetic energy be if the source is changed to Mg $K\alpha$?**

a) Planck's law is used to convert wavelength to the energy of the X-ray source in J and then convert J to eV (Note: 1 eV = 1.602×10^{-9} J, 1 nm = 10^{-9} m).

$$\nu = \frac{hc}{\lambda} = \frac{(6.626\times10^{-34}\,\text{J s})\times(3.0\times10^8\,\frac{\text{m}}{\text{s}})}{0.83393 \times 10^{-9}\text{m}} = 2.38\times10^{-16}\text{J} = 1487.9 \text{ eV}$$

b) Equation 14.5 is used to calculate the binding energy of the ejected electron in XPS:

$$E_b = h\nu - E_k - w = 1487.9 - 694 - 12.5 = 781 \text{ eV}$$

   c) The binding energy for a given element (Co 2p) remains the same (i.e., 781 eV) regardless of the change in X-ray source.

   d) As Eq. 14.6 implies, the kinetic energy in AES is independent of the incident source energy, whether it is Mg Kα or Al Kα, this kinetic energy of the Auger electron is still 694 eV.

6) **The current US EPA allowed maximum contaminant level (MCL) for α particles in drinking water is 15 pCi/L. Convert this into Bq/L and dpm/L.**

$$1\,\text{Curie} = 3.70 \times 10^{10}\,\text{Bq} = 2.22 \times 10^{12}\,\text{dpm}$$

$$1\frac{\text{pCi}}{\text{L}} = 1 \times 10^{-12}\frac{\text{Ci}}{\text{L}} \times \frac{3.70 \times 10^{10}\ \text{Bq}}{1\ \text{Ci}} = 0.037\frac{\text{Bq}}{\text{L}}$$

$$0.037\frac{\text{Bq}}{\text{L}} \times \frac{2.22 \times 10^{12}\ \text{dpm}}{3.70 \times 10^{10}\ \text{Bq}} = 2.22\frac{\text{dpm}}{\text{L}}$$

Thus, 1 pCi = 2.22 dpm, i.e., 1 pCi/L = 2.22 dpm/L

15 pCi/L = 15 × 2.22 = 33.3 dpm/L

Printed and bound by CPI Group (UK) Ltd, Croydon, CR0 4YY

08/10/2024

14570358-0001